Freiburger Empirische Forschung in der Mathematikdidaktik

Herausgegeben von
A. Eichler
Kassel, Deutschland

L. Holzäpfel
T. Leuders
K. Maaß
G. Wittmann

Freiburg, Deutschland

Die Freiburger Arbeitsgruppe am Institut für Mathematische Bildung (IMBF) verfolgt in ihrem Forschungsprogramm das Ziel, zur empirischen Fundierung der Mathematikdidaktik als Wissenschaft des Lernens und Lehrens von Mathematik beizutragen. In enger Vernetzung innerhalb der Disziplin und mit Bezugsdisziplinen wie der Pädagogischen Psychologie oder den Erziehungswissenschaften sowie charakterisiert durch eine integrative Forschungsmethodik sehen wir Forschung und Entwicklung stets im Zusammenhang mit der Qualifizierung von wissenschaftlichem Nachwuchs. Die vorliegende Reihe soll regelmäßig über die hierbei entstehenden Forschungsergebnisse berichten.

Herausgegeben von
Prof. Dr. Andreas Eichler
Universität Kassel

Prof. Dr. Lars Holzäpfel
Prof. Dr. Timo Leuders
Prof. Dr. Katja Maaß
Prof. Dr. Gerald Wittmann

Pädagogische Hochschule Freiburg, Deutschland

Carola Bernack-Schüler • Ralf Erens
Andreas Eichler • Timo Leuders (Eds.)

Views and Beliefs in Mathematics Education

Results of the 19th MAVI Conference

Springer Spektrum

Editors
Carola Bernack-Schüler
University of Education Freiburg
Germany

Andreas Eichler
University of Kassel
Germany

Ralf Erens
University of Education Freiburg
Germany

Timo Leuders
University of Education Freiburg
Germany

ISSN 2193-8164   ISSN 2193-8172 (electronic)
Freiburger Empirische Forschung in der Mathematikdidaktik
ISBN 978-3-658-09613-7   ISBN 978-3-658-09614-4 (eBook)
DOI 10.1007/978-3-658-09614-4

Library of Congress Control Number: 2015936040

Springer Spektrum
© Springer Fachmedien Wiesbaden 2015
This work is subject to copyright. All rights are reserved by the Publisher, whether the whole or part of the material is concerned, specifically the rights of translation, reprinting, reuse of illustrations, recitation, broadcasting, reproduction on microfilms or in any other physical way, and transmission or information storage and retrieval, electronic adaptation, computer software, or by similar or dissimilar methodology now known or hereafter developed.
The use of general descriptive names, registered names, trademarks, service marks, etc. in this publication does not imply, even in the absence of a specific statement, that such names are exempt from the relevant protective laws and regulations and therefore free for general use.
The publisher, the authors and the editors are safe to assume that the advice and information in this book are believed to be true and accurate at the date of publication. Neither the publisher nor the authors or the editors give a warranty, express or implied, with respect to the material contained herein or for any errors or omissions that may have been made.

Printed on acid-free paper

Springer Spektrum is a brand of Springer Fachmedien Wiesbaden
Springer Fachmedien Wiesbaden is part of Springer Science+Business Media
(www.springer.com)

Editorial

Ralf Erens und Carola Bernack-Schüler

Looking at venues of the various MAVI conferences, the 19th international conference on Mathematical Beliefs (MAVI) moved southwards into the heart of the Black Forest in southwestern Germany after it had been held in Talinn, Bochum and Helsinki in the last three years respectively. The 19th MAVI conference was organized by the University of Education in Freiburg from September 25th to September 28th, 2013. The current proceedings is/are published in the Freiburg Springer series of empirical research in mathematics education. We are grateful for the financial support of the University of Education Freiburg which enabled the publication of this conference issue

The founding fathers of these conferences, Erkki Pekhonen (Helsinki) und Günter Törner (Duisburg), initiated the MAVI group as a bilateral Finnish-German cooperation in 1995 which soon grew into an international community of researchers who met at yearly meetings e.g. in Pisa and Genova (Italy), Vienna and St. Wolfgang (Austria), Kristianstad and Gävle (Sweden) and Nikosia (Cyprus). The remarkable body of MAVI volumes of each meeting displays a variety of research papers on beliefs, attitudes and emotions in mathematics education.

A noteworthy characteristic of MAVI conferences is the lack of formal organization. Each participant enjoys an equal status and all accepted contributions are given the same time for presentation and discussion. According to the MAVI tradition, papers were submitted in advance and pass through a peer review procedure. For the 19th MAVI altogether 21 papers were presented and actively discussed by the participants.

The first section of this volume consists of six papers looking into teachers' beliefs. Their working contexts reach from pre-school (Sumpter, p.63) across out-of-field mathematics teaching (Bosse & Törner, p.10) to secondary school. The papers focus on mathematics in general but also on specific contents as curriculum reform (Berg et al., p.75) or beliefs about geometry (Girnat, p.xy). But also the belief change over 25 years in the Finnish teacher population was subject of research (Oksanen & Pehkonen, p.35) as well as conceptions about mathematics in different countries (Salo i Nevado et al., p.51).

The following four papers focus on belief research during teacher training and during the experiences as novice teachers (teacher trainees). The professional development of pre-service teachers is described either based on a longitudinal case study of a novice teacher (Palmér, p.131), taking into account external influences outside university (Ebbelind, p.119) or based on belief changes through a teacher training course (Bernack-Schüler et al., p.91). Furthermore the novice teachers' belief system concerning the teaching and learning of arithmetics was subject of research (Bräunling & Eichler, p.105).

The next three contributions deal with mathematics beliefs in the domain of technology. One paper is looking into pre-service teachers' beliefs concerning the teaching of mathematics with technology using metaphors (Portaankorva-Koivisto, p.135) but also the beliefs about the implementation and use of technology for the specific domain of calculus is described (Erens & Eichler, p.143). Moreover, Sundberg (p.169) glance at teachers' technological pedagogical content knowledge.

Another two papers look into beliefs related to problem solving and posing (Kontorovich, p.181; Papadopoulos, p.193). Beyond one theoretical framework from the history of mathematics is presented to investigate beliefs (Haapasalo & Zimmermann, p.207) and epistemological judgments about the certainty of mathematics knowledge were subject to research using a special approach in terms of interviews (Rott et al., p.235). Liljedahl & Andra (p.223) get a deep insight in students' interactions.

The different contributions address different topics and groups of students, teachers etc. The 19th MAVI conference added a variety or research perspectives to the international discussions of mathematics related beliefs and affect. With the feedback of the reviewers and the discussion in the conference the authors of this volume produced the existing contribution of a rich selection of research methods and may further enhance the discussion of MAVI topics in the future.

The editors
Carola Bernack-Schüler
Ralf Erens
Andreas Eichler
Timo Leuders

Contents

Editorial V

Teachers´ Beliefs

Marc Bosse & Günter Törner
Teacher Identity as a Theoretical Framework for Researching
Out-of-Field Teaching Mathematics Teachers 1

Boris Girnat
Teachers' Geometrical Paradigms as Central Curricular Beliefs in the
Context of Mathematical Worldviews and Goals of Education 15

Susanna Oksanen, Erkki Pehkonen & Markku S. Hannula
Changes in Finnish Teachers' Mathematical Beliefs and an Attempt to
Explain Them 27

*Laia Saló i Nevado, Päivi Portaankorva-Koivisto, Erkki Pehkonen,
Leonor Vara, Markku S. Hannula & Liisa Näveri*
Chilean and Finnish Teachers' Conceptions on Mathematics Teaching 43

Lovisa Sumpter
Preschool Teachers' Conceptions about Mathematics 55

Benita Berg, Kirsti Hemmi & Martin Karlberg
Support or Restriction: Swedish Primary School Teachers' Views
on Mathematics Curriculum Reform 67

Beliefs in Teacher Training and Novice Teachers' Beliefs

Carola Bernack-Schüler, Timo Leuders & Lars Holzäpfel
Understanding Pre-Service Teachers' Belief Change during a Problem Solving Course 81

Katinka Bräunling & Andreas Eichler
Teachers' Beliefs Systems Referring to the Teaching and Learning of Arithmetic 95

Andreas Ebbelind
We Think so, Me and My Mother – Considering External Participation Inside Teacher Training 109

Hanna Palmér
Primary School Teachers' Image of a Mathematics Teacher 121

Beliefs and Technology

Ralf Erens & Andreas Eichler
The Use of Technology in Calculus Classrooms – Beliefs of High School Teachers 133

Päivi Portaankorva-Koivisto
Mathematics Student Teachers' Metaphors for Technology in Teaching Mathematics 145

Maria Sundberg
A Study of Mathematics Teachers' Conceptions of Their Own Knowledge of Technological Pedagogical Content Knowledge (TPACK) 159

Beliefs and Problem Solving

Igor Kontorovich
Why Do Experts Pose Problems for Mathematics Competitions? 171

Ioannis Papadopoulos
Beliefs and Mathematical Reasoning during Problem Solving across Educational Levels 183

Further contributions

Lenni Haapasalo & Bernd Zimmermann
Investigating Mathematical Beliefs by Using a Framework from the History of Mathematics 197

Peter Liljedahl & Chiara Andra
Students' Gazes: New Insights into Student Interactions 213

Benjamin Rott, Timo Leuders & Elmar Stahl
Epistemological Judgments in Mathematics: An Interview Study Regarding the Certainty of Mathematical Knowledge 227

Authors 239

Teacher Identity as a Theoretical Framework for Researching Out-of-Field Teaching Mathematics Teachers

Marc Bosse, Günter Törner

University of Duisburg-Essen, Germany

Inhaltsverzeichnis

1 Introduction and Motivation 2
2 Teacher Identity as a Research Approach 3
    2.1 A Rough Survey of Existing Research 3
    2.2 Complex-Systemic Aspects of Teacher Identity 4
    2.3 Historic-Process-Oriented Aspects of Teacher Identity 5
    2.4 Narrative Aspects of Teacher Identity 6
3 From Teacher Identity to Subject-Related Teacher Identity 6
    3.1 Professional Competence of Mathematics Teachers 6
    3.2 Connecting Cognitive and Affective-Motivational Components 7
    3.3 Affective-Motivational Components as Shortcomings-Compensating Factors 8
4 Critical Reflection 9
    4.1 Challenges and Advantages of the Identity Approach 9
    4.2 Methodological Considerations towards Further Research Activities 10
5 Conclusions and Recommendations for Professional Development Programs 10
6 References 11

Abstract

This theoretical essay deals with teacher identity as an approach for studying out-of-field teaching mathematics teachers, i.e. teachers without formal education for teaching mathematics. It thinks about teacher identity as a unifying concept that is able to connect cognitive and affective-motivational perspectives. The connecting framework helps us to explain how out-of-field teaching in mathematics works: teachers' shortcomings in cognitive domains are indirectly compensated due to the prioritization-function of affective-motivational realms. Furthermore, other advantages of the identity approach are stated by referring to characteristics of identity concepts that can be found in literature. Finally, our conclusion leads to recommendations for designing in-service training programs for these teachers.

1 Introduction and Motivation

Some teachers have neither been formally educated as mathematics teachers at university nor in pre-service courses but actually teach mathematics in school. Following the terminology of Ingersoll (2001), we want to call this group of teachers *out-of-field teaching mathematics teachers*. In Germany – for example – we can observe this out-of-field teaching phenomenon both in primary schools and in secondary schools.

Only recently, the German *Institut zur Qualitätsentwicklung im Bildungswesen (IQB)* found out that in some federal states the percentage of out-of-field teaching mathematics teachers is about 50 % in grade 4 (Richter, D., Kuhl, P., Reimers, H., & Pant, H. A., 2012) and about 37 % in grade 9 (Richter, D., Kuhl, P., Haag, N., & Pant, H. A., 2013). Further, the authors of the quoted studies claim to show that there are significant differences in mathematics-related student achievement between students who were taught out-of-field and those who were not. Especially low-performing students achieved even worse, when educated by out-of-field teaching teachers.

Not only in Germany (Bosse & Törner, 2012; Törner & Törner, 2012) but also in other countries out-of-field teaching poses a challenge for teachers, teacher educators and politics (Crisan & Rodd, 2011; Dee & Cohodes, 2008; Hobbs, 2012; Ingersoll, 2001; Ingersoll & Curran, 2004; McConney & Price, 2009a; McConney & Price, 2009b; Rodd, 2012; Vale, 2010).

Having in mind that out-of-field teaching has negative influence on student achievement, it seems reasonable to cope with the phenomenon. Professional development programs – like those provided by the German National Centre for Mathematics Teacher Education (DZLM) – deal with this topic and offer in-service teacher education in order to support teachers in teaching mathematics. The crucial questions in this context are: What support do out-of-field teachers need? How can professional development programs help these teachers? And what has to be considered when designing in-service training courses? In order to answer these questions we have to understand the practical experiences made by those teachers. We have to get insight into their needs, into their self-conception, into their cognitive and affective problems and challenges, and – of course – into their relation to mathematics and to mathematics education.

Therefore, we need a theoretical framework in which we can analyze not only the phenomenon per se but the individuals who teach out-of-field. In the following paragraphs we will show that the theory of *teacher identity* is a useful approach to answer the crucial questions. We will proceed as follows:

First, we will explain the rather general term *identity* by referring to current research literature and by highlighting its benefits for our research purposes.

Second, we will connect the concept of *identity* to an established theoretical approach in mathematics teacher education: we will show how the identity concept can profitably enhance models for describing and analyzing teachers' professional competencies. The aim of this step is to develop a theoretical framework we want to call *subject-related teacher identity*. This framework helps both analyzing out-of-field teaching mathematics teachers and understanding how to and where to start when support shall be offered.

Third, we will critical reflect on the framework of *subject-related teacher identity*. Therefore, we will consider practical and methodological challenges that occur when researching out-of-field teaching mathematics teachers.

Fourth, we will make some concluding remarks about further steps in a possible research project. It has to be pondered whether, and if so, how the framework actually helps with designing professional development activities.

2 Teacher Identity as a Research Approach

2.1 A Rough Survey of Existing Research

Researching *Identity* is a broad field that is dealt with in many different educational and non-educational disciplines. In the last decades, the concept of identity has often changed depending on the theoretical perspective and depending on problems researchers want to work on. Grootenboer, Lowrie and Smith (2006) suggest that one can have three different views on the concept of identity: a psychological-developmental perspective, a social-cultural perspective and a poststructural perspective. Recently, researches in (mathematics) education have preferred social-cultural approaches (e.g. Boaler, 2002) like the theory of Communities of Practice by Wenger (2007). Apart from that, one can find studies in which teacher identity is seen in a poststructural framework (e.g. Walshaw, 2004). Other authors differ from the named trisection of perspectives. In this case, identity is seen as a narrative or at least in narrative contexts (Alsup, 2006; Sfard & Prusak, 2005).

On the one hand, each of these four perspectives implicates another research approach and another understanding of the identity concept. On the other hand, multiple perspectives (or facets of several single perspectives) are chosen and arranged for a specific research purpose. According to Grootenboer et al. (2006), the way the term identity is used in research contexts depends on the persuasions of the researcher.

Thus, it is difficult to provide an exact definition of the term *identity*. Some authors (e.g. Kelchtermans, 2009) prefer a quite narrow definition of identity by fixing static identity components, by framing their scope and by determining the meaning of the identity concept with a new term. Kelchtermans (2009) for example speaks about *self-understanding* instead of *identity*, because he wants to emphasize the teachers' conceptions of themselves as teachers.

Having in mind that out-of-field teaching of mathematics is an almost unexplored research field, we prefer a broader definition of *identity*. Therefore, we want to refer to Grootenboer et al. (2006) and Grootenboer and Zevenbergen (2008). They provide a "unifying" and "connecting" concept that can bring together multiple and interrelated elements which are relevant for out-of-field teaching mathematics teachers' professional activities, too. More precisely, the authors suggest using the term *identity* by framing beliefs, attitudes, emotions, cognitive capacity and life history. In this way, the identity concept lets us explore many different facets of out-of-field teaching mathematics teachers' activities, challenges and needs. Further, we cannot only have a look at cognitive but also on affective questions. With the help of this identity concept, the different identity facets are not detached anymore. They are brought together and allow us to gain a holistic view of our research objects.

2.2   Complex-Systemic Aspects of Teacher Identity

As a consequence of the broad definition, teacher identity becomes more complex. In fact, complexity appears in four ways:

First, the unifying concept of identity allows us to draw links to other relevant domains. Beauchamp and Thomas (2009) give an overview of such links and show that teacher identity is at least connected to activities of self-reflection, to the role of emotions, to agency, to stories and discourses, and to context. These domains are partly connected to each other. Thus, we can speak about identity as a systemic concept.

Second, we have to respect the situational character of identity (Beauchamp & Thomas, 2009; Grootenboer & Zevenbergen, 2007; Wenger, 2007). According to that, identity is something highly dynamic. It cannot be treated like a static entity you can have a look at without considering physical, social, institutional and affective contexts. If identity is context-dependent, we have to assume that there is something like a mathematical identity and a specific identity of out-of-field teaching mathematics teachers. Having said this, we should have in mind that teaching out-of-field is another context than teaching in-field and that *crossing* from one context to the other might have implications on many different levels (Hobbs, 2012).

Third, the dependence on context implies that identity comprises the existence of something like *sub-identities*. The identity of a *german out-of-field teaching primary mathematics teacher* belongs somehow or other to the identity of the *german primary teacher* that in turn belongs to the identity of the *german teacher* and so on. Such vertical identity hierarchies as well as horizontal identity overlapping – for example the identities of a mathematics teacher who also teaches geography – have to be considered.

Fourth, every teacher has two sub-identities that come along with special implications and interdependences: the professional and the personal identity (Alsup, 2006). In the context of our research purposes, this fact plays an important role in two ways: On the one hand, the out-of-field teaching teachers are professionals in teaching one (or more) subject(s) apart from mathematics. One can assume that the knowledge and the skills that are related to that (or these) subject(s) influence the personal identity. On the other hand, we can assume that a strong and developed professional identity in teaching mathematics is missing. However, these teachers encountered mathematics and teaching mathematics in their personal, non-professional life when they were students themselves. Thus, the personal identity might play a significant role when talking about the professional one.

2.3 *Historic-Process-Oriented Aspects of Teacher Identity*

In the identity concept, there is not only a dynamic momentum due to contextuality but also due to variability. Beauchamp and Thomas (2009) assemble different terms that are used in literature and that explain the process of identity shifting and reshaping (e.g. development, construction, formation, making, creation, building, architecture and so forth). Every term describes that identity is something that has to be built. The authors underline that identity development is not a process that eventually ends. Moreover, they say that identities are developed and re-developed constantly – which is obvious as the individual enters new contexts and makes new experiences constantly.

Also Sachs (2001) and Kelchtermans (2009) emphasize the importance of the dynamic nature of identity. In Kelchtermans' opinion, identity itself is an "ongoing process of making sense of one's experiences and their impact on the 'self'" (ibid., p. 261). Regarding the broad definition, talking about an individual's identity also means talking about life history (Grootenboer et al., 2006) and about the individual's biographic experiences (e.g. Alsup, 2006).

When using identity as an approach, the individual's past, present and future is considered. Bernstein (2000) distinguishes between the retroperspective and the properspective identity, having in mind that identity has developed from experiences in the past and will be developed from expectations and goals in the future. Also Sfard and Prusak (2005) consider the time-related variability of identity. They differentiate between the actual and the designated identity by explaining that there is a difference between actual experiences and those that are expected in the future.

2.4 Narrative Aspects of Teacher Identity

Stories and discourses are not only a way to express identity (Beauchamp & Thomas, 2009). Moreover, they are seen as a means of identity-making (ibid.; Alsup, 2006) or even as the identity itself (Sfard & Prusak, 2005). For the analysis of out-of-field teaching mathematics teachers it is mainly important, that narrative activities and communicational practice can support these teachers in developing a stronger mathematical identity (e.g. when consulting colleagues or attending an in-service course). In relation to this, Alsup (2006) ascribes narratives the power of transformation in thinking. Narratives become also relevant for methodological considerations, as they are able to open windows through which one can have a look at a facet of an individual's identity.

3 From Teacher Identity to Subject-Related Teacher Identity

3.1 Professional Competence of Mathematics Teachers

One option to analyze and to describe mathematics teachers' knowledge, skills and affective-motivational dispositions is the use of competence models. Especially empirical research deals with measuring teachers' competencies in order to model these with *competence profiles* (Blömeke, Kaiser, & Lehmann, 2010; Blömeke, Suhl, & Döhrmann, 2012). For example, the model that is used in the TEDS-M study (ibid.) contains both cognitive (PCK, CK, PK) and affective-motivational (beliefs, motivations, self-regulation) domains. Of course, we could list other competence models for modeling mathematics teachers' competencies.

3.2 Connecting Cognitive and Affective-Motivational Components

Having a look at each of these models, the question is left open how these two domains are connected to each other. The authors of such models try to describe teachers' cognitive and affective competencies as precisely as possible by dividing them into sub-competencies that are sometimes divided into sub-sub-competencies again. Once a research field is defined and a corresponding (sub-)-competence is located in the model, other competencies are faded out (above all affective-motivational competencies when someone has chosen a cognitive (sub)competence). At the best, a study is extensive enough to research many different competencies and sub-competencies. That has to be the necessary condition if someone wants to get a holistic picture of a teacher, his or her skills, needs, and shortcomings.

The unifying concept of *identity* as described above gives us the chance to build a theoretical framework in which cognitive and affective-motivational components are already connected to each other. The framework does not claim to be able to define fixed competence fields but can, however, provide a holistic picture. Moreover, the theory makes it possible to explore the same research objects based on both cognitive *and* affective perspectives.

When researching out-of-field teaching mathematics teachers, it stands to reason that the cognitive focus is on mathematics and teaching mathematics. Using the terminology of the established competence models, the theory of identity is able to relate affective-motivational competencies to any item dedicated to PCK and CK. Having this in mind, we do not have to care about asking ourselves if we want to analyze a sub-competence of PCK, if we want to explore a CK related question exclusively or if we are just interested in a perspective shedding light on a teacher's beliefs.

Let us have a look at an example. If we want to analyze how out-of-field teaching mathematics teachers cope with the Theorem of Pythagoras, how they implement the theorem in teaching contexts and how a professional development program can support these teachers in doing that, we can examine this against the background of identity very well.

One can observe how they encountered the theorem in professional and/or personal domains and ask whether, and if so, how this plays a role in present teaching (*life history* aspect of identity). Further, one can ask whether, and if so, how they have actually learned to give a fruitful introduction to this topic in lessons. Supporting colleagues might play an important role when answering this question (*contextuality* and *variability* of identity). Apart from that, one can research which proof of the theorem the teacher uses in teaching and why he or she favors it (connections to facets of *self-regulation* and *self-understanding*). It is also an interesting opportunity to ask about likes and dislikes around the theorem and its possible proofs (*emotional* aspects) and about personal and professional experiences, expectations and views towards it (aspects of *beliefs*, *attitudes* and *biography*). Of course, one could find many other practical approaches.

Every identity-related item of this example is somehow or other linked to the subject (CK and/or PCK). Therefore, we want to call the theory *subject-related teacher identity*.

3.3 *Affective-Motivational Components as Shortcomings-Compensating Factors*

When dealing with out-of-field teaching mathematics teachers, we have to assume that they have shortcomings in PCK and CK, as they were not educated in these domains (Bosse & Törner, 2012). Schoenfeld (2011) suggests that the elements of the affective-motivational domain ("orientations") have a function of prioritizing cognitive and other resources. If cognitive resources are missing, then other resources are consulted (e.g. colleagues, textbooks, own experiences in life history). What other resources are used for teaching and how the process of choosing resources looks like, is – according to Schoenfeld – initiated by the affective-motivational domain (cf. ibid, p. 30). From this, one can conclude that out-of-field teaching teachers' shortcomings in the cognitive domain are indirectly compensated by affective-motivational components.

In this respect, the theory of *subject-related teacher identity* – as combining both components – helps to clarify basics and mechanisms of compensating. Therefore, it helps to understand how teachers teach out-of-field and how teachers can be supported.

4 Critical Reflection

4.1 Challenges and Advantages of the Identity Approach

Argument of ethics and appreciation: In our opinion an analysis that investigates shortcomings of out-of-field teaching mathematics teachers exclusively is not an option. Some of these teachers are highly motivated and enthusiastic to teach mathematics out-of-field. Thus, their work has to be appreciated instead of highlighting the deficits. A study that concentrates on shortcomings would ignore that these teachers already compensate drawbacks by their own methods and strategies. The identity framework makes these approaches visible and considers both shortcomings and the individual ways to cope with them.

Argument of dissolving uncertainty: It seems to us that analyzing competence profiles carries a certain risk. Similar to Heisenberg's uncertainty principle, investigating a competence object to close – for example in the context of a sub-sub-competence field – leads to loosing information about the object itself. The identity approach is a more holistic way that allows us to research many different facets of out-of-field teaching mathematics teachers without the need to be perfectly precise on subatomic-like levels. Nevertheless, one gets a broad picture of teaching and possible starting points for supporting these teachers.

Argument of cognitive-empowerment: The cognitive and the affective domain are related to each other. The identity approach respects this due to its unifying character. Separating cognitive and affective components, as it is often done in competence-oriented approaches, limits the opportunities to understand interdependencies that are especially relevant for out-of-field teaching teachers. That has to be underlined since affective components are able to empower cognitive ones (see a multitude of findings in the wake of emotion, motivation and belief research).

Argument of model learning: Not only teachers but also students own a mathematical identity (Boaler, 2002). In order to avoid that students acquire negative attitudes towards mathematics, teachers should develop a fruitful mathematical identity themselves. Bandura (1977) showed that students learn from role models. If an out-of-field teaching mathematics teacher is not able to develop an appropriate identity, his or her students will not do either (see Boaler, 2002; Grootenboer & Zevenbergen, 2008).

Argument of practice: The starting points for the study as well as the practical implementation of the findings are related to questions of teaching in practice. The theoretical framework considers this: Since identity and practical experiences can be made accessible due to narratives, the approach provides ways for involving practical relevance.

4.2 Methodological Considerations towards Further Research Activities

While out-of-field teaching mathematics teachers often refuse to participate in our studies due to affective reasons (fear, shame, embarrassment), it is difficult to undertake a broad quantitative study. Further, it is often unknown who actually teaches out-of-field, as there are almost no statistics about the phenomenon. The identity concept provides a theoretical framework that can also be used for doing reasonable qualitative research on a small group of teachers.

Despite of the group size, the identity approach lets us achieve findings that provide a holistic picture. A qualitative analysis of different and manifold facets is possible – and not only an investigation of a specific competence-field. The identity approach helps us to research *beliefs*, contexts of *self-image* and *sense of self*, *motivations* and *emotions*. Every facet can be projected into mathematics-related realms, since the identity approach is a unifying concept.

Besides, the identity concept provides narrative approaches that are especially appropriate for qualitative interviews and narrative data collection. To this effect, a useful research method is – for example – asking out-of-field teaching mathematics teachers to write short essays about the topic "mathematics and me".

5 Conclusions and Recommendations for Professional Development Programs

Sachs (2005) claims, that teachers' "professional identity [...] stands at the core of the teaching profession. It provides a framework for teachers to construct their own ideas of 'how to be', 'how to act' and 'how to understand' their work" (p. 15). If we project this into mathematics and mathematics education as explained above, we are able to comprehend how to support out-of-field teaching mathematics teachers. Subject-related teacher identity as a theoretical framework provides starting-points for both research and intervention due to professional development. Further, it allows us to focus not only on shortcomings in cognitive knowledge but to consider affective-motivational dispositions. As affective-motivational components are responsible for prioritizing knowledge and other resources (cf. Schoenfeld, 2011), we can explain how out-of-field teaching works in spite of shortcomings in CK and PCK.

A professional development program for out-of-field teaching teachers should not only spend time on fostering subject-related cognitive competencies. Of course, this is necessary and important; but in addition to that, in-service training courses should have an eye on the teachers' subject-related identity. We are convinced that being able to explain out-of-field teaching in such a holistic way leads to a better understanding for designing effective and successful professional development.

6 References

Alsup, J. (2006). Teacher identity discourses: Negotiating personal and professional spaces. Mahwah, NJ: Lawrence Erlbaum Associates.

Bandura, A. (1977). Social learning theory. Prentice-Hall series in social learning theory. Englewood Cliffs, N.J: Prentice Hall.

Beauchamp, C., & Thomas, L. (2009). Understanding teacher identity: An overview of issues in the literature and implications for teacher education. Cambridge Journal of Education, 39(2), 175–189.

Bernstein, B. B. (2000). Pedagogy, symbolic control, and identity: Theory, research, critique (Rev. ed.). Critical perspectives series. Lanham, Md: Rowman & Littlefield Publishers.

Blömeke, S., Kaiser, G., & Lehmann, R. (Eds.). (2010). TEDS-M 2008: Professionelle Kompetenz und Lerngelegenheiten angehender Primarstufenlehrkräfte im internationalen Vergleich. Münster et al.: Waxmann.

Blömeke, S., Suhl, U., & Döhrmann, M. (2012). Zusammenfügen was zusammengehört: Kompetenzprofile am Ende der Lehrerausbildung im internationalen Vergleich. Zeitschrift für Pädagogik, 58(4), 422–440.

Boaler, J. (2002). The development of disciplinary relationships: knowledge, practice, and identity in classrooms. For The Leaning of Mathematics, 22(1), 42–47.

Bosse, B., & Törner, G. (2012). Out-of-field Teaching Mathematics Teachers and the Ambivalent Role of Beliefs – A First Report from Interviews. Unpublished Paper presented at the 18th MAVI Mathematical Views conference, Helsinki, September 2012.

Crisan, C., & Rodd, M. (2011). Teachers of mathematics to mathematics teachers: a TDA Mathematics Development Programme for Teachers. In C. Smith (Ed.), Proceedings of the British Society for Research into Learning Mathematics (pp. 29–34). Retrieved from http://www.bsrlm.org.uk/IPs/ip31-3/BSRLM-IP-31-3-06.pdf

Dee, T. S., & Cohodes, S. R. (2008). Out-of-Field Teachers and Student Achievement: Evidence from Matched-Pairs Comparisons. Public Finance Review, 36(1), 7–32.

Grootenboer, P., Lowrie, T., & Smith, T. (2006). Researching identity in mathematics education: The lay of the land. In P. Grootenboer, R. Zevenbergen, & M. Chinnappan (Eds.), Identities cultures and learning spaces. Proceedings of the 29th annual conference of the Mathematics Education Research Group of Australasia (Vol. 2, pp. 612–615). Canberra, Australia: MERGA.

Grootenboer, P., & Zevenbergen, R. (2007). Identity and mathematics: Towards a theory of agency in coming to learn mathematics. In J. Watson & K. Beswick (Eds.), Mathematics: Essential Research, Essential Practice. Proceedings of the 30th annual conference of the Mathematics Education Research Group of Australasia (Vol. 1, pp. 335–344). Adelaide: MERGA.

Grootenboer, P., & Zevenbergen, R. (2008). Identity as a Lens to Understand Learning Mathematics: Developing a Model. In M. Goos, R. Brown, & K. Makar (Eds.), Navigating currents and charting directions. Proceedings of the 31st Annual Conference of the Mathematics Education Research Group of Australasia (Vol. 1, pp. 243–249).

Hobbs, L. (2012). Teaching 'out-of-field' as a Boundary-Crossing Event: Factors Shaping Teacher Identity. International Journal of Science and Mathematics Education.

Ingersoll, R. M. (2001). Rejoinder: Misunderstanding the problem of out-of-field teaching. Educational Researcher, 30, 21–22.

Ingersoll, R. M., & Curran, B. K. (2004). Out-of-Field Teaching: The Great Obstacle to Meeting the "Highly Qualified" Teacher Challange. Retrieved from http://www.nga.org/files/live/sites/NGA/files/pdf/0408HQTEACHER.pdf

Kelchtermans, G. (2009). Who we am in how we teach is the message: Self-understanding, vulnerability and reflection. Teachers and Teaching, 15(2), 257–272.

McConney, A., & Price, A. (2009a). *An Assessment of the Phenomenon of "Teaching-Out-of-Field" in WA Schools*. Perth, Australia: Western Australian College of Teaching.

McConney, A., & Price, A. (2009b). *Teaching Out-of-Field in Western Australia, 34*(6), 86–100.

Richter, D., Kuhl, P., Haag, N., & Pant, H. A. (2013). Aspekte der Aus- und Fortbildung von Mathematik- und Naturwissenschaftslehrkräften im Ländervergleich. In H. A. Pant, P. Stanat, U. Schroeders, A. Roppelt, T. Siegle, & C. Pöhlmann (Eds.), *IQB-Ländervergleich 2012. Mathematische und naturwissenschaftliche Kompetenzen am Ende der Sekundarstufe I* (pp. 367–390). Münster/New York/München/Berlin: Waxmann.

Richter, D., Kuhl, P., Reimers, H., & Pant, H. A. (2012). Aspekte der Aus- und Fortbildung von Lehrkräften in der Primarstufe. In P. Stanat, H. A. Pant, K. Böhme, & D. Richter (Eds.), *Kompetenzen von Schülerinnen und Schülern am Ende der vierten Jahrgangsstufe in den Fächern Deutsch und Mathematik. Ergebnisse des IQB-Ländervergleichs 2011* (pp. 237-250). Münster: Waxmann.

Rodd, M. (2012). In-service courses for teachers of mathematics: identity, equity and mathematics. Retrieved from http://www.ioe.ac.uk/about/documents/About_Overview/Strand_1_-_Rodd_M(2).pdf

Sachs, J. (2001). Teacher professional identity: competing discourses, competing outcomes. Journal of Educational Policy, 16(2), 149–161.

Sachs, J. (2005). Teacher education and the development of professional identity: Learning to be a teacher. In P. Denicolo & M. Kompf (Eds.), Connecting policy and practice. Challenges for teaching and learning in schools and universities (pp. 5–21). London, New York: Routledge.

Schoenfeld, A. H. (2011). How we think: A theory of goal-oriented decision making and its educational applications. New York: Routledge.

Sfard, A., & Prusak, A. (2005). Telling Identities: In Search of an Analytic Tool for Investigating Learning as a Culturally Shaped Activity. Educational Researcher, 34(4), 14–22.

Törner, G., & Törner, A. (2012). Underqualified Math Teachers or Out-of-Field-Teaching in Mathematics - A Neglectable Field of Action? In W. Blum, R. Borromeo Ferri, & K. Maaß (Eds.), Mathematikunterricht im Kontext von Realität, Kultur und Lehrerprofessionalität. Festschrift für Gabriele Kaiser (pp. 196–206). Wiesbaden: Vieweg+Teubner Verlag.

Vale, C. (2010). Supporting "out-of-field" teachers of secondary mathematics. The Australian Mathematics Teacher, 66(1), 17–24.

Walshaw, M. (2004). Preservice mathematics teaching in the context of schools: An exploration into the constitution of identity. Journal of Mathematics Teacher Education, 7, 63–86.

Wenger, E. (2007). Communities of Practice: learning, meaning, and identity (1. paperback ed., 15. print.). Learning in Doing. Cambridge et al.: Cambridge Univ. Press.

Teachers' Geometrical Paradigms as Central Curricular Beliefs in the Context of Mathematical Worldviews and Goals of Education

Boris Girnat

University of Applied Sciences and Arts Northwestern Switzerland School of Teacher Education, Basel, Switzerland

Content

1 Interest of Research and Background of the Study 16
2 Theoretical Background 16
    2.1   Geometrical Working Spaces 16
    2.2   Mathematical Worldviews...17
    2.3   Goals of Education..19
3 Settings of the Study and Methodological Background 20
4 Empirical Findings 20
5 Conclusions 23
6 References 26

Abstract

This article presents some results of a qualitative study on secondary teachers' beliefs, reconstructed as so-called individual curricula, a concept to represent a teacher's argumentative connections between his choice of content, methods, and goals of education. Within these individual curricula, two archetypes are figured out that are supposed to be oppositional in three dimensions: in the use of Geometrical Working Spaces in classroom teaching, in the general mathematical worldview, and in the choice of goals of education a teacher intends to achieve by teaching elementary geometry. The first archetype is characterised by deductive standards, a static view on mathematics, and expert-oriented goals of education; the second one is more empirical, dynamic, and guided by pragmatic goals of education.

1 Interest of Research and Background of the Study

This article presents some results of a qualitative interview study concerning secondary school teachers' individual curricula on teaching elementary geometry. The core framework is based on the concept of *individual curricula* (Eichler, 2007) which are used to describe the part of a teacher's *beliefs system* (cf. Philipp, 2007) that contains argumentative connections between content, methods, and goals of education and has a similar function as a written curriculum (cf. Stein, Remillard & Smith, 2007), especially the task to justify the choice of contents and teaching methods against to the background of a teacher's individual goals of education.

After reconstructing nine individual curricula out of in-depth interviews, the study was faced to the problem to compare and to categorise the findings. Since an individual curriculum – even restricted to teaching elementary geometry – is a "holistic" conception, it is not sufficient to use just one framework for a categorisation, e. g. just a purely geometrical one; rather it is advisable to use discriminations on three typical levels of a curriculum: the level of goals of educations, the geometrical level, and the geometrical aspect seen in a broader context of general beliefs of the "nature" of mathematics. To do so, three background theories were combined, namely the *theory of Geometrical Working Spaces* (Kuzniak, 2006), a classification of *goals of education* (Graumann, 1993) and a framework to analyse general understandings of mathematics, called the theory of *mathematical worldviews* (Grigutsch, Raatz, & Törner, 1998). Insofar, the central research question of this study is as follows: How can individual curricula on teaching elementary geometry be classified based on these three levels and are there any systematic connections between them? It will be argued that the answer is positive and that it is possible to identify two archetypes of systematic connections between these levels and that each of the nine teachers can be attached to one of the two archetypes.

2 Theoretical Background

Before we can start to describe the study and its method, it is necessary to make some remarks on the three theoretical backgrounds used for the classification.

2.1 Geometrical Working Spaces

The framework of Geometrical Working Spaces (GWS) is based on the idea that three *geometrical paradigms* are relevant to the history and philosophy of elementary geometry which are fundamentally different in ontological, epistemological, and practical assertions (Houdement & Kuzniak, 2001). The classification consists of three entries which are named and explained as follows:

1) Geometry I or G1 (Natural Geometry): Geometry is regarded as an *empirical discipline* which refers to physical objects. To "proof" or to refute conjectures, both *argumentations and experiments* are allowed. The basic foundations of arguments are not axioms, but propositions derived from empirical observations. The standards of arguments are typically not as "sophisticated" as in mathematical proofs, but close to ordinary language argumentations used in everyday life.

2) Geometry II or G2 (Natural Axiomatic Geometry): Geometry is seen as an *axiomatic theory*. The axioms are supposed to refer to the real world and, therefore, to describe physical figures and objects (with some idealisations); but to proof or to reject propositions, no empirical arguments are allowed. Only *deductive* conclusions based on the axioms are permitted.

3) Geometry III or G3 (Formalist Axiomatic Geometry): Geometry is seen as a *formal axiomatic theory*, and no connection to the real world is intended.

G3 is more or less restricted to university level, whereas G1 and G2 are the paradigms that play a role at secondary school. Against to the background of geometrical paradigms, a *pupil's* Geometrical Working Space can be described as his individual (conscious or unconscious) selection of aspects of one or more geometrical paradigms he uses when being confronted to geometrical tasks, concepts, figures, and problems (Kuzniak, 2006). This approach was extended to analyse *teachers'* standards of teaching geometry (Girnat, 2009). In this case, the teacher's GWS is not necessarily his own working space, but the working space he demands from his pupils to use. Insofar, the teacher's GWS expresses what type of geometrical paradigms he wants to see as predominant in his lessons on geometry.

According to Houdement & Kuzniak (2001), the main problem on teaching geometry consists in the fact that a written curriculum normally intended the use of G2, whereas pupils often adhere on G1. Girnat (2009) pointed out that the teachers' response to this problem is quite diverse: Some of teachers try to implement G2 standards as their intended GWS, but some prefer to teach geometry on a G1 level, partly intentionally to avoid pupils being demanded more than "appropriate", partly unintentionally since G1 is their own understanding of geometry.

2.2 Mathematical Worldviews

A mathematical worldview can be explained as a *beliefs system* (cf. Philipp, 2007) which a person, especially a teacher, holds for true and "essential" in all parts of mathematics. We follow an approach of Grigutsch, Raatz, & Törner (1998) who suggest a classification of mathematical worldviews by four aspects:

1) *Formalistic aspect*: Mathematics is seen as a formalistic language whose concepts are introduced by definitions and whose theorems are derived by deduction from basic axioms.

2) *Schematic aspect*: Mathematics is seen as a pool of rules and algorithms which enables a person to solve mathematical problem by following these rules and algorithms (like recipes, i. e. not necessarily by understanding their backgrounds).

3) *Dynamic aspect*: Mathematics is seen as a field of creativity in which everyone can try to invent his own concepts and rules to solve mathematical problems or problems including a mathematical part. The opposite is called the *static aspect*, which means: Mathematics is seen as a bound of theories whose concepts, axioms, and theorems are fixed and unchangeable; and doing mathematics means reproducing these theories and to applying them correctly.

4) *Applied-oriented aspect*: Mathematics is seen as practically useful and as a powerful tool to handle challenges occurring in everyone's professional and everyday life.

Grigutsch, Raatz, & Törner (1998) undertook a representative study among secondary school teachers (N=400) to reveal correlations between the four aspect of their mathematical worldviews. Their results are presented in Figure 1.

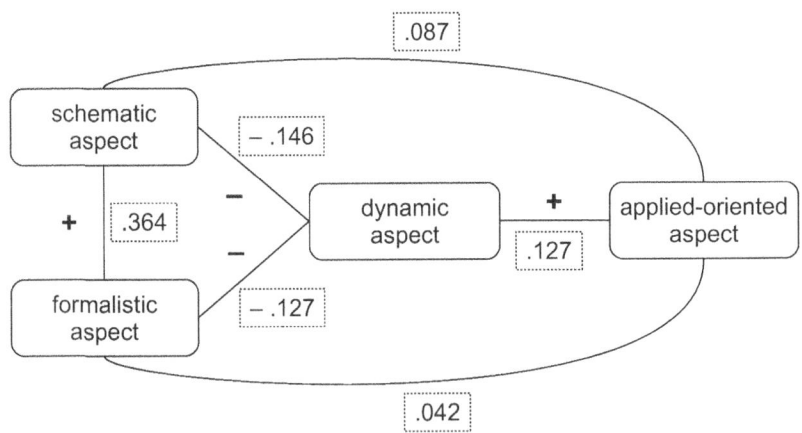

Figure 1 Correlations between the aspects of mathematical worldviews

The correlations are low at all hands, but nevertheless, Grigutsch, Raatz, and Törner conclude that there are two clusters, namely a cluster which consist of the schematic and the formalistic aspect and an opposed cluster which is formed by the dynamic and the applied-oriented one. This hypothesis is taken into account in our study. More precisely, there are two questions of interest: 1) Can similar affinities between the four mathematical worldviews observed in the study; 2) and are there connections between a teacher's mathematical worldview and his choice of the GWS he demands his pupils to use.

2.3 Goals of Education

There are many approaches to classify goals of mathematics education. To analyse our teachers' statements, we choose a model that depends on two steps of discrimination: At first, we distinguish if a teacher wants to make his pupils achieve competencies which are mainly *specifically mathematical* or if he is interested in using mathematical education to acquaint his pupils with goals of education that are *more general* than mathematical ones. Let us call the first point of view *expert education*, the latter one *general education*. In case of expert education, the specific goals are given by the teacher's understanding of mathematics and are related to his mathematical worldview or (nor specially) to his geometrical paradigm. In case of general education, we use a framework of Graumann (1993) to distinguish between five dimensions of general education:

1) *Pragmatic dimension*: Mathematics education should be perceived as useful to solve practical and technical problems.

2) *Enlightenment dimension*: Mathematics education should foster an understanding of the world including its historical, cultural, and philosophical backgrounds.

3) *Social dimension*: Mathematics education should strengthen the pupils' competencies to cooperate, to communicate, and to accept responsibility.

4) *Individual dimension*: Mathematics education should enhance each pupil's own abilities and interests.

5) *Reflective dimension*: Mathematics education should sensitise the pupils to the limits, boundaries, and fallacies of mathematical methods.

The choice of this framework is founded in need of a conception which is decidedly not restricted to mathematical education (as e. g. the widespread frameworks of mathematical competencies would be), but which is in principle applicable to every school subject.

3 Settings of the Study and Methodological Background

The study was carried out at higher-level secondary schools by interviewing nine teachers about their *individual curricula* (cf. Eichler, 2007) of teaching geometry. All these teachers studied mathematics at university on a level comparable to a master of science without or with just a minor contingent of pedagogy or didactics of mathematics. They gained their certificate necessary to be employed as secondary school teachers in a practically oriented second step of training after their studies at university. They were chosen randomly from nine different schools.

The method to reconstruct individual curricula from interview transcripts is based on a qualitative approach, called *dialogue-hermeneutics technique*, which was invented to expose argumentative relations within belief systems (Groeben & Scheele, 2000). In our case, the argumentative relations in question are means-ends relations between contents, teaching methods, and goals of education. The methods is based on a way to represent such connections graphically: If a teachers utters a sentence like "I do a lot of problem solving to enhance my pupils' intellectual skills", the aim "enhancing intellectual skill" is placed in a tree diagram on a higher level and the mean "by doing a lot of problem solving" is subordinated to this aim on a lower level. After the interviewer has compiled such a diagram in the hermeneutic stage of the method, his end-product is given to the teacher to check if he can approve the interviewer's proposal or if he insists on revising the diagram to display his arguments correctly. This is the dialogical part of the method. In fig. 2 and 3, diagrams derived by this technique are shown which are condensed to a very abstract structure.

The interviews are prepared along the principle to give as little input as possible. Therefore, the questions normally are very open like "Could you describe your lessons on geometry?" and typically followed by questions which are supposed to reveal the teacher's aims like "Why do you prefer this content, this teaching method and so on?" or "Why do you do this and not for example this alternative?"

4 Empirical Findings

Seven of the teachers who took part of our study can be classified as proponents of G2, two of them as exponents of G1. The latter ones are called Ernest and Henry. To quote a typical passage which can be used to classify a teacher's GWS, we choose some statements of proponents of G2 first and some of Ernest's and Henry's later:

Ian: Geometry definitely is a well-ordered system, if you follow Euclid's "Elements". It is a prototypic example of an axiomatic theory. Unfolding this theory at school is impossible, but on a local level, it is a very important to make pupils argue precisely, to deduce from premises, to make them search for proofs or to retrace proofs at least. […] As a mathematician, I have to observe that pupils are not simply convinced by empirical observations.

Gertrude: It's the central point of mathematics to argue logically and to show the pupils how logical chains of proofs are made.

Fredric: It is important to me that my students switch to an abstract level, practise pure geometry. In order to do so, applications, concrete figures, measuring and so on are rather obstacles than aids. These are no valid methods.

Dorothy: The beauty of mathematics is the fact that everything is logical and dignified. […] Everywhere else, there are approximations, but not in mathematics. There is everything in this status it has ideally to be in. [It is important for the pupils] to recognise that there are ideal things and objects in mathematics and that, in reality, they are similar, but not equal.

Ernest: A theory has got its place at university. […] Theoretical deliberations only make sense at school – like in my lessons –, if they are useful to solve practical problems. I mean authentic problems that come from the pupils' everyday life. Otherwise, a theory is deathlike. […] A proof is something conflicting. Normally, you prove theorems at school just because someone said that's the task mathematician have to do; and that's brainless, I think. For me, argumentation is more a social phenomenon to convince each other, to discuss a problem together, to help each other. That's the social aspect. […] It would be nice if we had a problem and everyone would propose different concepts, definitions, and we would try how far we can.

Henry: Proofs are of minor interest. The task is to make theorems plausible, e. g. by cutting out figures and laying them onto each other, and you can observe if they match each other; and we take this as a proof. […] DGS [dynamic geometry software] is a very useful tool. You can prove a lot by it. For example, in case of Thales, you can put a third point on this semi-circle and you can pull it hither and thither; and you will notice that the angle equals 90 degrees all the time. And so, you have proven that there are always 90 degrees, and you did it convincingly. […]At the end of the day, it's not necessary to be exact; it is necessary that pupils can solve their tasks. That doesn't have to be exact. It depends on the context. And later, it will be important to the pupils to solve problems. It won't be important to solve them elegantly or in the manner they have learnt at school; it will be just important that they are willing to face the problem and that they will come to a suitable solution, an estimation, an approximation anyhow.

Like in these quote, the crucial distinction between G1 and G2 is found in the role of justification and in the perception of geometrical objects of being empirical or non-empirical, "idealistic" ones. The first passages stress the importance of deductive proofs, whereas Henry and Ernest are willing to accept empirical observations as arguments. Beside that fundamental difference, you can observe some remarks on the topics we want to combine with the teachers' geometrical paradigms: 1) Ian stresses his role "as a mathematician", and Dorothy pointed out some ontological and epistemological aspects she regards as typical for mathematics. She concludes that they are "therefore" also important to her pupils. Henry, on contrary, does not accept the argument that an aspect of mathematics has to be part of mathematics education just because it is typical for mathematics as a (scientific) discipline. Besides these quotes, which can only illustrate the findings, the proponents of G2 tend to emphasise an expert education in mathematics and consider it as their task to familiarise their pupils to a scientifically oriented perception of mathematics. 2) The proponents of G1 stress the pragmatic benefits of geometry and the importance of mathematical education for the pupils' future life. They seem to be willing to adjust mathematical standards of exactness and justification to the circumstances which are given by a real world problem mathematics is a part of. Insofar, the exponents of G1 appear to be applied-oriented, whereas the proponents of G2 seem to fear confusions between mathematical and empirical standards of justification, if the practical use comes to close to geometry. 3) Especially Ernest emphasises the social dimension of education and seems to represent a more dynamic conception of mathematics which includes creating new concepts and exploring them in the contexts of realistic problems. 4) The observation that G1 proponents are willing to adjust mathematical standards of exactness to practical needs may indicate that they perceive mathematics as a "tool box" in a schematic manner. But this is unclear.

Let us regard some further excerpts of the interviews to search for connections to other aspects of goals of educations and mathematical worldviews among G2 proponents, since until now, they seem to be just focussed on expert education:

Gertrude: Besides proof abilities, problem solving is in fact the most important thing I want to convey in my lessons on geometry. [...] I want activate my pupils to deliberate on their own and to overcome the habit "Now, we handle ten task using the recipe xy".

Ian: Problem solving is a sort of intelligent exercises. You have to remember former content, and you have to use it actively. Thinking, I mean, intellectual abilities are trained by problem solving; and to be successful, you have to be fit in mathematical basics, and you have to train them.

Frederick: Mathematics education is brain callisthenics. In other school subjects, you can learn something different, but in my lessons, you will do brain callisthenics. [...] Proofs are important. You can't accept an assertion just because someone told you that it's true. You have to scrutinise everything.

Alan: I think, in mathematics education, pupils can learn to think, to structure, to solve problems. [...] And beside this, I want to give my pupils an insight how the ancient Greek did it. They were very ambitious. They didn't just want to know how something was, they wanted to found why it was as it was. Normally, the pupils don't want to be inferior to them.

Christian: All the tasks provided at school are fabricated. I have no concern to provide a task that is fabricated. The pupils will accept it, and they can learn geometry even if the task is fabricated. It's the same thing in every school subject, and mathematics education has not to apologise for this fact. […] Knowing the basic principles precisely and drawing conclusions from these principles without calling them into question, I think that's something you can learn in mathematics education, and not in other school subjects.

In case of G1, we can primarily find the goal to show the practical side of mathematics and to prepare pupils to their further life, as sketched above. These quotes, on contrary, shall illustrate some typical statements in which proponents of G2 express goals of education that lie beyond subject-specific aspects: 1) It is noticeable that they disdain authentic real-world problems. They seem to regard realistic tasks only as tools to learn mathematics, and not as a subject interesting of its own. Insofar, they attach little value to the pragmatic dimension of education, but stress the enlightenment and reflective dimension, promoting formal and intellectual abilities. 2) Beside proofs, problem solving task seem to be the main focus of the G2 proponents. 3) For both proofs and problem solving, they seem to regard it as necessary to possess a broad and active knowledge of basic principles which are standardised and not committed to subjective creations. Insofar, a more static view of mathematics seems to be a precondition to teach geometry in the sense of G2 and to achieve goals of education that are seemingly linked to this kind of teaching.

5 Conclusions

Considering the few passages quoted here and the few teachers interviewed, it seems daring to draw general conclusions. But since it is one of the main tasks of qualitative studies not to make representative assertions, but to generate archetypes which can be tested representatively afterwards, we propose two archetypes that represent systematic connections between the three layers of our classification, namely goals of educations, geometrical paradigms, and mathematical worldviews.

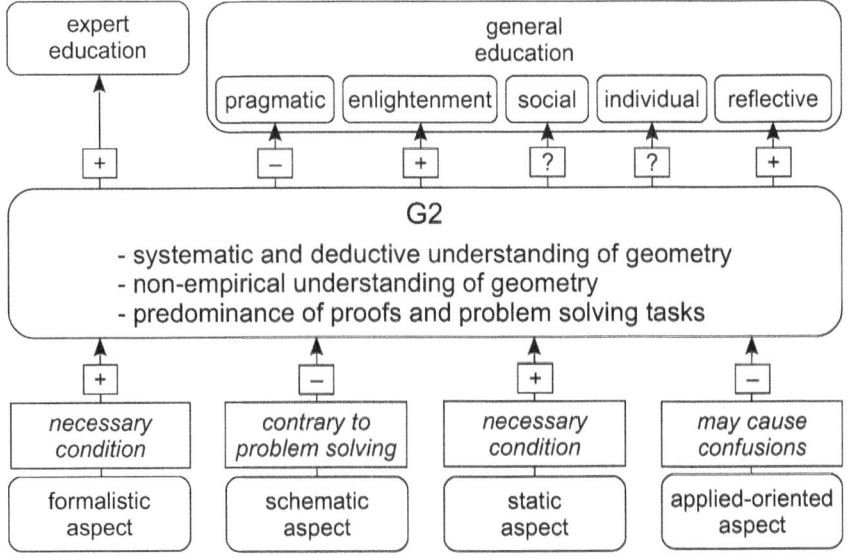

Figure 2   G2 in the context of mathematical worldview and goals of education

In fig. 2, we try to sketch the connections that are supposed to be typical for the G2 GWS. A plus sign indicates that the G2 teachers affirm a specific goal of education (at the top of the diagram) or a specific mean (at the bottom). The minus signs denote refusals. In case of a question mark, neither an affirmative nor a dismissive statement can be found. If a quantitative extension of this study was carried out, the plus and minus signs would indicate the hypotheses that a positive or rather a negative correlation should be observable. The italic phrases display typical reasons "in a nutshell" the teachers use to justify their approvals and rejections.

Overall, the formalistic and the static aspect of mathematics seem to be a precondition to implement G2 standards. Applications may be "too empirical" and could cause a conflict with the non-empirical standards of justification. It seems a "natural" way to extend the G2 approach to subject-specific goals of education, especially to an expert education and, as far as general education is concerned, to rather intellectual and cultural aspects than to pragmatic ones.

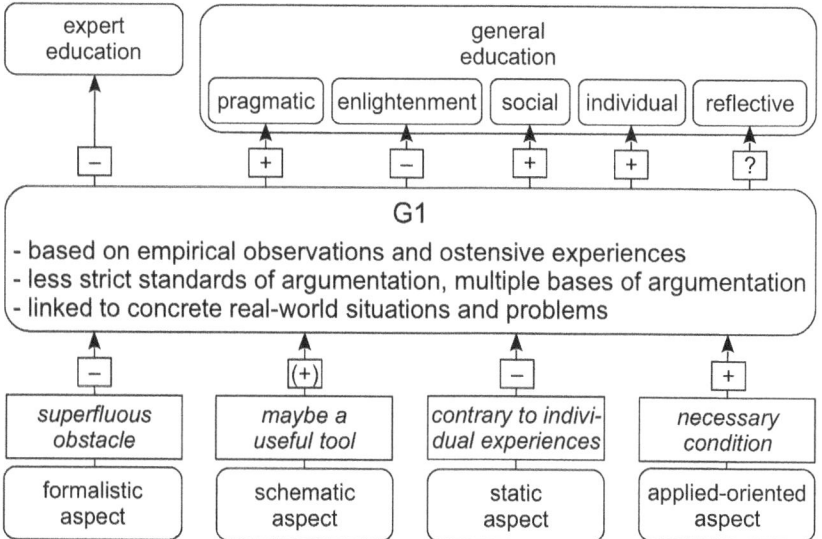

Figure 3 G1 in the context of mathematical worldview and goals of education

As visualised in fig. 3, an archetypical G1 curriculum is supposed to look quite oppositional compared to a G2 one: Realistic problems are necessary conditions for an empirical understanding of geometry, and from there, a "natural" way leads to pragmatic goals of education. Formalistic aspects are not as important as for a G2 concept and rather obstacles; and a static understanding of mathematics seems to be opposed to individual experiences.

The results of this study may be useful in two respects: 1) Qualitative studies primarily support conceptual work and hypothesis generation. In this study, we propose the two concepts of a G1 and G2 archetypical curriculum. These suggestions can be the initial points for a representative study on this issue. In this case, the two concepts GWS and "dimensions of general education" has to be operationalised); and it has to test if the connections proposed by plus and minus signs in the archetypes can be confirmed or not.

2) If these archetypes were affirmed by a representative study, it would be possible to get a deeper explanation of mathematical worldviews, since in this case, the teachers' GWS would be a "hidden variable" that could explain systematic connections "behind" correlations as displayed in fig. 1. Maybe, it is possible to revise or to clarify some of the aspect expressed there. For example, the highest correlation observed, the one between the formalistic and the schematic aspect, is astonishing concerning the G2 teachers' statements that they prefer problem solving and that they want to reduce the amount of schematic tasks. A possible

explanation could be that they insist on routine task as a precondition of problem solving, but that they do not see schematic tasks as valuable on their own. Insofar, the connection to the archetypical curricula could be the basis to formulate the aspects of mathematical worldviews more precisely and more usefully to collect data on teachers' curricular background of planning their lessons on geometry.

6 References

Eichler, A. (2007): Individual curricula – Teachers' beliefs concerning stochastics instruction. International Electronical Journal of Mathematics Education 2(3). http://www.iejme.com/.

Girnat, B. (2009): Ontological Beliefs and Their Impact on Teaching Elementary Geometry. Proceedings of the 33rd IGPME conference, Thessaloniki, vol. 3, pp. 89–96.

Graumann, G. (1993): Die Rolle des Mathematikunterrichts im Bildungsauftrag der Schule. Pädagogische Welt, 5, pp. 194–199.

Grigutsch, S., Raatz, U., &Törner, G. (1998): Einstellungen gegenüber Mathematik bei Mathematiklehrern. Journal für Mathematik-Didaktik (JMD), 19(1), pp. 3–45.

Groeben, N., & Scheele, B. (2000). Dialogue-Hermeneutic Method and the "Research Program Subjective Theories". Forum: Qualitative Social Research 1(2), Art. 10, http://nbn-resolving.de/urn:nbn:de:0114-fqs0002105.

Houdement, C. & Kuzniak, A. (2001): Elementary Geometry Split into Different Geometrical Paradigms. Proceedings of CERME 3. Bellaria (Web).

Kuzniak, A. (2006), Paradigmes et espaces de travail géométriques. Éléments d'un cadre théorique pour l'enseignement et la formation des enseignants en géométrie, Canadian Journal of Science, Mathematics and Technology Education, 6(2), 167–188.

Philipp, R. A. (2007): Mathematics Teachers' Beliefs and Affect. In F. K. Lester (Ed.), Second Handbook of Research on Mathematics Teaching and Learning (pp. 257-315). Charlotte: Information Age Publishing.

Stein, M. K., Remillard, J., & Smith, M. S. (2007). How curriculum influences pupil learning. In F. Lester (Ed.), Second Handbook of research on mathematics teaching and learning (pp. 319–369). Charlotte: Information Age Publishing.

Changes in Finnish Teachers' Mathematical Beliefs and an Attempt to Explain Them

Susanna Oksanen, Erkki Pehkonen, Markku S. Hannula
University of Helsinki

Content

1 Introduction	28
2 Theoretical framework	28
2.1 Mathematics teachers' beliefs	29
2.2 Mathematics teaching in Finland	29
3 Research question	30
4 Methodology	30
4.1 Procedure	31
4.2 Sample	31
4.3 Analyses	32
5 Results of the questionnaires	32
8 Discussion	35
8.1 The frames of the Finnish mathematics teaching	35
8.2 Change factors in the Finnish system	36
8.3 Changes around mathematics curricula	37
8.4 Increase of didactical know-how	38
8.5 An attempt to explain why such changes	38
9 Conclusions	38
10 References	39

Abstract

This study reveals what kind of changes have occurred in Finnish mathematics teachers' beliefs during 1987-2012 and some reasons for the change. The information gained with two questionnaires and their analysis show what kind of beliefs about mathematics and its teaching teachers had 25 years ago, and how they have changed. This paper also presents the most important influential factors on the development of Finnish mathematics teaching within four decades, the years 1970–2010. The main point of the paper is to reveal the recent rapid development of the Finnish school system and society.

Key words: mathematics, teachers' beliefs, teacher change

1 Introduction

Teachers' beliefs play a significant role in the planning and implementation of their teaching. They affect on what will be taught, how it is taught, and therefore, what is learned in the classroom (Andrews & Hatch, 2000). Beliefs shape how teachers think and feel about mathematics and its teaching and learning. As teachers' beliefs affect their teaching, it is important to recognize those beliefs. Changing teachers' practices will depend on changing their beliefs, and changing beliefs will lead to change in practices (Lerman, 2002).

In the last decades the content and ways of teaching mathematics have changed. Based on the results of the survey called TALIS (Teaching and Learning International Survey), it is evident that teaching style has a big influence on pupils' performance (OECD, 2009).

Pehkonen & Lepmann (1995) conducted a survey on Finnish mathematics teachers' beliefs about mathematics and mathematics teaching in 1987-1988 using Dionné framework (1984). It introduces three perceptions of mathematics: traditional view (mathematics is limited to calculations and following rules), formalist view (stresses rigorous logic, proofs and exact use of language), and constructivist view (the pupil comes first and the emphasis is on pupil-centered learning methods and intuition).

Oksanen & Hannula (2013) have repeated the survey in 2010-2011 and report similarities and differences between these two measurements. They expected that the changes in curriculum and the overall teaching philosophy in teacher education would be reflected also in teachers' beliefs.

2 Theoretical framework

Pehkonen and Törner (1998) summarized that an individual's mathematical beliefs are compound of his subjective, experience-based, implicit knowledge on mathematics and its teaching and learning. The spectrum of an individual's beliefs is very large, and its components influence each other.

We base our construction of beliefs and referring terminology on the article of Op't Eynde, De Corte, and Verschaffel (2002), who have strived for making a synthesis regarding previous belief researches. In the paper Op't Eynde and others (2002) define mathematical beliefs to be implicitly or explicitly held subjective conceptions people hold to be true, that influence their mathematical learning and problem solving.

Beliefs may have a knowledge-type nature, (e.g. view of mathematics: "mathematics learning is independent on gender"), the truthfulness of which can be discussed in social interaction, or volitional nature (individual and subjective; such as "It is important to me to provide good experiences with mathematics"). The latter kind of beliefs' validity can never be judged socially with any "scientific criteria". Beliefs, as such, are subjective, something that an individual believes to be true, no matter whether the others agree or disagree. (Op't Eynde & al, 2002).

2.1 Mathematics teachers' beliefs

Lerman (2002) underlines that there is a strong link between beliefs and practices: changing teachers' practices will depend on changing their beliefs, and changing beliefs will lead to change in practices. Teacher change consists changes in teachers' classroom behavior but also in the very art of teaching.

The importance of reflection in changing teachers' beliefs has also being recognized. According to Tobin (1990) reflective thinking about teaching can change the teaching behavior and actions. In addition to reflection, teachers' ability to attend to students' understanding of mathematics and to base given instructions on what and how students are thinking is also important.

2.2 Mathematics teaching in Finland

Finland has been successful in the four last international PISA comparison studies (2000, 2003, 2006 and 2009). Recently the teacher educators in Helsinki wrote a book on the Finnish educational system, especially in the case of mathematics and science (cf. Pehkonen, Ahtee & Lavonen 2007). This book tries to describe what kind of changes have occurred in the Finnish school system, especially in mathematics education. Since the target group in the PISA studies was 15-year-olds, the paper at hand concentrates on the upper level of the comprehensive school, i.e. on grades 7–9 (13–16 years old pupils).

Mathematics education in Finland has been changing rapidly within the last 60 years, but the change has been the most rapid within the four last decades (cf. Pehkonen 2008). Apparently a part of changes is due to the school reform in the

1970's. Until the end of the 1960's mathematics education in secondary school in Finland was relatively steady and stable. The recent structure of the Finnish school system is described e.g. in the published paper (Lavonen 2006) and in the book (Pehkonen & al. 2007).

In Finland the traditional or conventional teaching methods have been prevalent in mathematics teaching. Teacher educators have been trying to enrich and to change these traditional methods with alternative, more pupil-centered methods (cf. Pehkonen & Rossi 2007). The traditional teachers emphasized basic teaching methods and extensive drilling, while the innovative teachers focused more on student thinking and deeper learning. (Kupari, 1996). This change of teaching methods has got more energy through the emergence of constructivism, in Finland in the 1980's.

In the paper Oksanen & Hannula (2013), it was used the same standardized factor scales as Loogma, Ruus, Talts & Poom-Valickis in 2009: traditional beliefs (the teaching is first and foremost the direct transmission of knowledge from the teacher to the pupil) and constructivist beliefs (the main emphasize is on the development of thinking and the understanding of the causal connections).

3 Research question

What kind of changes have occurred in Finnish mathematics teachers' beliefs during 1987-2012 and some reasons for the change?

4 Methodology

In 1987–88 the study on Finnish teachers' beliefs was carried out for comparing Estonian and Finnish teachers of mathematical understanding (Pehkonen & Lepmann, 1995). In 2009 a new cross-cultural survey of mathematics teachers' beliefs was initiated. In this article a study that is a part of this cross-cultural survey of mathematics teachers is presented from the perspective of the comparison. The study focuses on mathematics teachers of lower secondary school (grades seven to nine). The teacher beliefs are categorized according to Dionné (1984) into three groups: traditional, formalist and constructivist.

Instrument in 1987-1988. The questionnaire included 54 structured items about different situations in mathematics teaching originating from the research project "Open Tasks in Mathematics (Pehkonen & Zimmermann, 1990). The teachers were asked to rate their views within these statements on a five point Likert scale. Thirty-five of these items were successfully classified into the three dimensions by Dionné (Pehkonen & Lepmann, 1995).

Instrument in 2011-2012. A questionnaire with 77 statements was built in connection with an international NorBa study (Nordic-Baltic Comparative Research in Mathematics Education). The questionnaire included seven modules, one of which included 24 'Dionné items' from the questionnaire 1987-1988. We removed from our analysis two items that were considered unreliable for comparison because they had changed in their wording. Theoretical background, development and structure of the questionnaire as well as the sample items for first three modules are described more thoroughly in the previous papers (Lepik & Pipere, 2011; Hannula, Lepik, Pipere & Tuohilampi, 2012).

4.1 Procedure

Procedure in 1987-1988. One part of the sample consisted of teachers on in-service courses (N=52), and the other group of teachers were reached by a postal questionnaire (N=34).

Procedure in 2011-2012. Informative letters and E-mails were sent to schools all over Finland inviting mathematics teachers to participate in the polling. Teachers who wished to participate in the polling filled in applications and sent them back to the university, or used an electronic form to inform about the willingness (N=94).

4.2 Sample

Sample in 1987-1988. The respondents were 86 Finnish mathematics teachers with different teaching experiences and ages. The youngest teacher was 27 years old and the oldest 57 years old. The average age was 41 years.

Sample in 2011-2012. The respondents were 94 Finnish mathematics teachers teaching grades 7-9 from different regions of Finland with different teaching experiences and ages. The sample size in 2010 was 52 teachers (the questionnaire was sent to 35 schools) and in 2011-2012 42 teachers (the questionnaire was sent to 71 schools). The average age of respondents was 41 with range of 25 to 61 years old. The average duration of teaching experience of respondents was 14,5 with range of 1 to 35 years.

4.3 Analyses

In both surveys we had 22 identical items. For the older data we used the means and standard deviations that were reported in the end of the research report from Pehkonen & Lepmann (1995). Because we could not confirm the equal variance across samples, we used the Welch's t-test to compare the changes in teachers' beliefs.

5 Results of the questionnaires

The statistically significant differences in 0,001-0,05 level between teachers' beliefs in 1987-1988 and 2011-2012 appeared in 45% of statements (10 statements out of 22). In the formalist view, statistically significant differences appeared in 1 out of 7 statements (14%), in the traditional view in 4 out of 7 statements (57%) and in the constructivist view in 5 out of 8 statements (63%).

In the following tables, the mean results of every statement in every category are presented, and the statistically significant differences are marked with * (0,05 level), ** (0,01 level) and *** (0,001 level).

Item number in the NorBa-questionnaire (2011-2012) is presented first. Item number in the earlier questionnaire (1987-1988) is in brackets. Likert-scale in the NorBa-questionnaire goes from 1=Fully disagree to 5=Fully agree and in the former questionnaire from 1=Fully agree to 1= Fully disagree. Therefore, the older results are scaled to respond NorBa-study's results.

Table 1 The formalist view

Item number	6 Item wording	1987-88 Mean and SD		2011-12 Mean and SD	
20 (35)	Mathematics teaching should emphasize logical reasoning	4,31	0,80	4, 38	0,779
12 (19)	In teaching, one should proceed systematically above all	4,03	0,94	3.83	0,872
1 (4)	One has to pay attention to the exact use of language (e.g. one should distinguish between an angle and the magnitude of an angle, between a decimal number and a decimal notation)	3, 86	1,15	3,70	0,953
11 (17) ***	In particular, the use of mathematical symbols should be practiced	2,76	1,17	3,41	0,888
18 (33)	Abstraction practice should be stressed in mathematics	3,06	0,96	3,02	0,766

		1987-88		2011-12	
4 (7)	Working with exact proof forms is an essential objective of mathematics teaching	2,20	0,99	2,48	1,022
8 (12)	The irrationality of the number √2 has to be proved	1,92	1,10	1,84	0,934

As expected, the constructivist way of teaching was in year 2012 supported most strongly (mean=3,72). The formalistic way (mean=3,23) and the traditional way of teaching (mean=3,17) were supported the least. The directions in the changes are negative in traditional and formalistic view. The changes in constructivist view go to both directions.

When taking the three most agreed statements in the year 2012, the beliefs of Finnish mathematics teachers can be described as follows: Mathematics teaching should emphasize logical reasoning, as often as possible; pupils should work using concrete materials; above all the teacher should try to get pupils involved in intensive discussions.

Table 2   The traditional view

Item number	Item wording	1987-88 Mean and SD		2011-12 Mean and SD	
13 (24)*	The learning of central computing techniques (e.g. applying formulas) must be stressed	4,01	0,91	3,71	0,867
6 (10)***	In mathematics teaching, one has to practice much above all	4,34	0,85	3,87	0,981
19 (34)*	Above all mathematical knowledge, such as facts and results, should be taught	2,86	1,14	2,48	0,940
2 (5)***	In a mathematics lesson, there should be more emphasis on the practicing phase than on the introductory and explanatory phase	4,21	0,86	3,48	0,974
17 (31)	As often as possible such routine tasks should be solved where the use of the known procedure will surely lead to the result	2,83	1,11	3,11	0,914
14 (26)	Pupils should above all get the right answer when solving tasks	2,65	1,14	2,41	1,055
16 (29)	A pupil need not necessarily understand each reasoning and procedure	3,20	1,04	3,14	1,054

It is obvious that the teachers emphasize the importance of logical reasoning and exact use of language and systematizing. But sometimes teaching should be implemented, for example, as project-oriented (beyond subject limits), instead of learning only the central computing techniques and solving routine tasks. In the traditional view statistically significant differences were found in four out of seven statements. The support of teaching facts and practice a lot of exercises have decreased, and the support to understanding increased.

The importance to use concrete materials has increased. Also the support on using social forms of learning, such as project work and class discussion, has increased. At the same time the independent thinking via problem posing and problem solving has lost some of its popularity amongst teachers.

Table 3   The constructivist view

Item number	7 Item wording	1987-88 Mean		2011-12 Mean	
21 (38)	Pupils should develop as many different ways as possible of finding solutions, and in teaching they should be discussed	3,95	0,95	3,88	0,845
22 (40)***	Pupils should formulate tasks and questions themselves, and then work on them	4,33	0,80	3,56	0,926
24 (47)***	As often as possible, the teacher should deal with tasks in which pupils have to think first and for which it is not enough to merely use calculation procedures	4,43	0,75	3,71	0,854
15 (28)	Above all the teacher should try to get pupils involved in intensive discussions	3,84	0,89	3,89	0,853
10 (16)*	As often as possible, pupils should work using concrete materials (e.g. cardboard models)	3,66	1,08	3,96	0,908
5 (9)*	Sometimes teaching should be realized as project-oriented (beyond subject limits), and prerequisites for it should be created. (An example of the project: to buy and maintain an aquarium.)	3,42	1,14	3,80	0,891

| 9 (14) | In mathematics teaching, learning games should be used | 3,67 | 1,06 | 3,87 | 0,875 |
| 3 (6)** | Mathematics has to be taught as an open system that will develop via hypotheses and *cul-de-sacs* | 3,52 | 1,08 | 3,12 | 0,931 |

8   Discussion

Here we discuss the most important changes in the Finnish school mathematics during the years 1970–2010. A lot of activities are still left outside, in order to deliver a coherent picture to the reader. Unfortunately many references are in Finnish, since they are written for our teachers, administrators and policy makers, but the titles of the publication are translated into English, in order to a reader is offered a flavor of its contents.

8.1   The frames of the Finnish mathematics teaching

In the beginning of the 1980's the dominant understanding of learning was a behaviorist one. The main idea of learning, also in mathematics, was repetition and practice. This understanding of learning can be read also in the textbooks of the time period in question.

Board of General Education began at the end of the 1980's to publish small booklets on the new understanding on learning that was based on constructivism (Lehtinen 1989). Among these booklets, there was also a discussion book on mathematics education (Halinen & al. 1991) in that a vision for the 1990's was created.

Departments of teacher education are aiming to change the traditional habit of teaching. Already in the 1980's Finnish teacher in-service education concentrated much on the delivering of new methods of school teaching. There were in-service courses for mathematics teachers on using learning games, on problem solving, on developing creativity, on using computers and calculators, and on constructive geometry teaching.

The amount of mathematics lessons in the upper level of the comprehensive school (grades 7–9) is in Finland one of the smallest in the whole world (UNESCO 1986): During the period of the two curricula (1985–1994 and 1994–2004), there was only 3 lessons mathematics pro week in each grade of the upper level of the comprehensive school (i.e. grades 7–9); today there is one additional weekly lesson for grade 9. Therefore, teachers are not very eager to use their teaching time to anything else, but to concentrate on the basic mathematics, during their few mathematics lessons.

8.2 Change factors in the Finnish system

When we start to look for explanations for the changes observed, one big change is the "new math" movement that was implemented also in Finland in the 1970's. It was called "new math", since teaching of structural mathematics was there in the paramount position, and teaching set theory was in the very center. In the Nordic countries the new math reform was implemented about ten years later than in the United States. In the Nordic co-operation, the committee report on renovation of mathematics education in the Nordic countries was elaborated and published at the end of the 1960's (Anon.). The report gave common directions for the renovation of mathematics education.

But at the same time a school reform was planned and implemented in Finland. The aim was to develop a comprehensive school instead of the existing parallel school system. The committee report for the curriculum in the comprehensive school was published in the beginning of the 1970's (Anon.). The parliament decided that in Finland the change to the uniform school system will be implemented during the years 1970–77. The reform meant that the civic school and intermediate school were emerged into a unified comprehensive school of grades 1–9. The reform began in 1970 from the North (Lapland) and moved little by little southwards; it reached finally in 1977 Helsinki and its surroundings.

Thus the two big reforms – the new math reform and the comprehensive school reform – were mixed with each other. Since in the comprehensive school the aim is to teach the whole age cohort together, this had influences also in the contents of mathematics education. The mathematics curricula of the civic school and intermediate school were emerged together. Problems rising from the teaching of the whole age cohort in mathematics together were tried to control with level courses (streaming) in the upper grades of the comprehensive school (grades 7–9). In the middle of the 1970's, the Board of General Education accepted that the curriculum for the comprehensive school (Anon.) was too demanding, and they published a suggestion for the so-called basic curriculum (Anon.). Therefore, the curriculum was rewritten in the beginning of 1980's (Anon.). A detailed description of this phase of curriculum reform is given e.g. in the published paper Pehkonen (1983).

8.3 Changes around mathematics curricula

At the end of the 1970's, citizens were not satisfied with the results of mathematics instruction in the comprehensive school. Especially the abstract "new math" was criticized, since people were afraid that children don't learn enough basic calculations. As in the States, also in Finland the movement "Back to Basics" was raising its head. Teachers wanted to enrich abstract (pure) school mathematics with different applications (Leino & Norlamo 1980).

Since the level courses were totally removed in 1985, we had from the middle of the 1980's heterogeneous groups in mathematics. And within each teaching group pupils' knowledge level varies a lot. In order to manage to teach in the heterogeneous groups, teachers and textbooks moved in the 1980's to the so-called "task didactics", since the only functioning solution seemed to be to differentiate with tasks. This movement was accelerated by the Nordic co-operational initiative for individualization in mathematics instruction (Anon.).

Work on new curricula began already in the beginning of the 1990's. But in the Board of General Education there was an evolution – it and the Board of Vocational Education were emerged and reorganized into a new National Board of Education. At the same time, the curriculum strategy was changed: it was decided to free schools from "the chains of curricula". National Board of Education published only the framework curriculum (Anon.) within which communes and schools were obligated to write their own curricula.

In 1996 teaching of mathematics and science in Finland got a new push forward, when the government decided to give special attention to the level of mathematics and science teaching. Ministry of Education published a six-year development program named National Joint Action (Heinonen 1996), and allotted six million Finn marks (about one million euros) to implement it. A revision of the Action was published three years later (Anon.), and the final report after six years (Anon.). National Joint Action produced many courses and activities, but failed in some main points, as in diminishing gender differences in learning results, and in increasing interest in studying mathematics and science. The main part of the results can be read also in another report (Anon.).

The development in school mathematics in the 2000 was no more so hectic. In 2004 the latest curriculum for the comprehensive school was published (Anon.). In that version, some of the obligations concerning the curriculum were taken away from teachers, in that sense the curriculum was transformed to more traditional than the earlier ones. Now there are initiatives for the rewriting of the curriculum, e.g. the new share of lessons for each subject is given and the groups for writing the subject matter part of the curriculum are named. It is planned that new curriculum will be published in 2016 (Anon.).

8.4 Increase of didactical know-how

In Finland, teacher education was reorganized and moved into the universities in 1974. At the same time, new positions on mathematics education were established in the universities; in some positions there was also a clear demand on research besides teaching.

About ten years later in 1983, teachers and researchers in mathematics and science education organized themselves into an association on research of teaching in mathematics and science education (Malinen & Kupari 2003). All this development has led in the 1990's to the situation that Finland has a clear leading position within the Nordic countries in mathematics education and its research, according to an outside evaluator (cf. Johansson 1994).

8.5 An attempt to explain why such changes

International comparison studies show that Finnish mathematics teaching has been implemented optimal to the resources in a long run. Success in the international comparison studies, as TIMSS and PISA (Välijärvi & al. 2012), was confirmed with the results of the following PISA rounds. Finland is today clearly in a leading position in the world, when measured with the PISA indicators, and in the TIMSS it was also clearly above the average.

If we measure the amount of pupils' knowledge, the Finns will be losers compared with other countries, since our weekly number of mathematics lessons is the smallest in Europe (UNESCO 1986). The small amount of weekly lessons does not offer an opportunity to learn mathematics at large in school. Therefore, school teaching has been concentrated more or less around calculation, and mathematical understanding seems to be very poor. About ten years ago the second author evaluated the state-of-art of Finnish school mathematics in a published paper Pehkonen (2001). In the paper, the fact was emphasized that there are lacks in the level of mathematical understanding of both school pupils and teacher students (also Merenluoto & Pehkonen 2002).

9 Conclusions

During the last two decades, researchers around the world have paid more and more attention to mathematics learning as a process. For example, Wilson and Cooney (2002) pointed that students learn mathematics most effectively when they construct meanings for themselves, rather than simply being told. A constructivist approach to teaching helps students to create these meanings and to learn. Learning is interactive and student-centered.

Nowadays the constructivist teaching approach is commonly in use. For example TALIS results (OECD, 2009) showed that in most countries endorsement of constructivist beliefs (beliefs regarding constructivist teaching approach) is stronger than that of traditional beliefs (beliefs regarding traditional teaching approach). This applied especially the Scandinavian countries.

In Finland much work has been done during more than three decades, in order to help teachers to change their teaching style from teacher-centered to pupil-centered. National Board of Education also aimed this via publishing a guide for implementation of mathematics teaching (Seppälä 1994) within the new framework curriculum (Anon.). For example, the focus of teacher in-service training in the 1980's was in different pupil activities, as learning games, problem solving and outdoor activities. Also written materials on the use of alternative teaching methods were published (e.g. Pehkonen & Rossi 2007); some of these are described also in the report Ahtee & Pehkonen (1994). In the 2000's one could state that in Finnish school mathematics, we have moved from the "task didactics" of the 1980's further to pupil activities.

10 References

Ahtee, M. & Pehkonen, E. (eds.) (1994). Constructivist Viewpoints for School Teaching and Learning in Mathematics and Science. University of Helsinki. Department of Teacher Education. Research Report 131.

Andrews, P., & Hatch, G. (2000). A comparison of Hungarian and English teachers' conceptions of mathematics and its teaching. Educational Studies in Mathematics, 43, 31–64.

Dionné, J. (1984). The perception of mathematics among elementary school teachers. In J. Moser (Ed.), Proceedings of the 6th Annual Meeting of the North American Chapter of the International Group for the Psychology of Mathematics Education (pp.223–228). Madison (WI): University of Wisconsin.

Heinonen, O.-P. (1996). Finnish know-how in mathematics and natural sciences in 2002. National joint action. Koulutus- ja tiedepolitiikan julkaisusarja 38. Helsinki: Opetusministeriö.

Hannula, M.S., Lepik, M., Pipere, A., & Tuohilampi, L. (2012). Mathematics teachers' beliefs in Estonia, Latvia and Finland. A paper presented in the Eight Congress of European Research in Mathematics Education (CERME 8).

Johansson, B. (1994). Nordisk matematikdidaktisk forskning idag. In: Matematik¬didaktik i Norden (red. O. Björkqvist & L. Finne), 1–44. Rapporter från Pedagogiska fakulteten vid Åbo Akademi, nr. 8/1994.

Kupari, P. (1996). Changes in teachers' beliefs of mathematics teaching and learning. In G. Törner (Ed.), Current state of research on mathematical beliefs II. Proceedings of the 2nd MAVI Workshop. Gerhard-Mercator-University, Duisburg, March 8-11, 1996. Schriftenreihe des Fachbereichs Mathematik (pp. 25-31). Duisburg: Gerhard-Mercator-University.

Lehtinen, E. (1989). Vallitsevan tiedonkäsityksen ilmeneminen koulun käytännöissä [Manifestation of the dominant conception on knowledge in school practices]. Kouluhallituksen julkaisuja Nro 18. Helsinki: Valtion painatuskeskus.

Leino, J. & Norlamo, P. (1980). Matematiikan kouluopetuksen sisältö tekniikan ja talouselämän näkökulmasta [The content of school mathematics from the viewpoint of technique and economy]. Matemaattisten aineiden opettajien liitto MAOL. Tutkimuksia N:o 1.

Lepik, M., & Pipere, A. (2011). Baltic-Nordic comparative study on mathematics teachers' beliefs: Designing research instrument to describe the beliefs and practices of mathematics teachers. Acta Paedagogica Vilnensia, 27,115-123.

Lerman, S. (2002). Situating Research on Mathematics Teachers' Beliefs and on Change. In G. C. Leder, E. Pehkonen, & G. Törner (Eds.), Beliefs: A Hidden Variable in Mathematics education? (pp. 233–243). Kluwer Academic Publishers, Netherlands.

Loogma, K., Ruus, V.-R., Talts, L., & Poom-Valickis, K. (2009). Õpetaja professionaalsus ning tõhusama õpetamis- ja õppimiskeskkonna loomine. OECD rahvusvahelise õpetamise ja õppimise uuringu TALIS tulemused. [Teacher Professionalism and Creating Effective Teaching and Learning Environment ts. Results from TALIS: Teaching and Learning International Survey of OEACD: In Estonian]. Tallinna Ülikooli Haridusuuringute Keskus. Available online at: http://www.hm.ee/index.php?048181 *est

Malinen, P. & Kupari, P. (2003). Miten kognitiivisista prosesseista kehiteltiin konstruktivismia [How constructivism was developed from cognitive processes]. Matematiikan ja luonnontieteiden opetuksen tutkimusseura. Koulutuksen tutkimuslaitos: Jyväskylän yliopisto.

Merenluoto, K. & Pehkonen, E. (2002). Elementary teacher students' mathematical understanding explained via conceptual change. In: Proceedings of the PME-NA XXIV (eds. Mewborne, D., Sztajn, P., White, D.Y., Wiegel, H. G., Bryant, R. L. & Nooney, K.), 1936–1939. Columbus (OH): ERIC

Oksanen & Hannula (2013). Changes in Finnish mathematics teachers' beliefs during 1987-2012. In: Proceedings of the PME conference in Kiel (ed. A. Heinze). A research report proposal submitted.

Pehkonen, E. (1983). Entwicklung des Geometrieunterrichts in Finnland innerhalb der letzten zwanzig Jahre. Mathematica didactica 6 (3/4), 117-128.

Pehkonen, E. (1998). On the concept "Mathematical belief". In E. Pehkonen & G. Törner (eds.) The State-of-Art in Mathematics-Related Belief Research; Results of the MAVI activities. Research report 184, (pp. 37–72). Department of teacher education.University of Helsinki.

Pehkonen, E. (2001). Mitä on matematiikka ja miten sitä osataan koulussa [What is mathematics and how is it ruled in school]. Arkhimedes 3/2001, 14–17.

Pehkonen, E. (2008). Some background factors for the Finnish PISA results in mathematics. Mediterranean Journal for Research in Mathematics Education, Vol. 7 (1), 51–62.

Pehkonen, E., Ahtee, M. & Lavonen, J. (Eds.) (2007). How Finns Learn Mathematics and Science. Rotterdam / Taipei: Sense Publishers.

Pehkonen, E. & Lepmann, L. (1995). Vergleich der Auffassungen von Lehrern über den Mathematikunterricht in Estland und Finnland. University of Helsinki. Department of Teacher Education. Research Report 139.

Pehkonen, E. & Rossi, M. (2007). Some alternative teaching methods in mathematics. In: E. Pehkonen, M. Ahtee & J. Lavonen (Eds.), How Finns learn mathematics and science, 141–152. Rotterdam / Taipei: Sense Publishers.

Pehkonen, E. & Zimmermann, B. (1990). Problem Fields in Mathematics Teaching. Part 1: Theorerical Background and Research Design. University of Helsinki. Department of Teacher Education. Research Report 86.

OECD. (2009). Creating Effective Teaching and Learning Environments: First Results from TALIS. Paris: OECD Publishing.

Op'T Eynde, P., de Corte, E., & Verschaffel, L. (2002). Framing students' mathematics-related beliefs. In G. C. Leder, E. Pehkonen, & G. Törner (Eds.), Beliefs: A Hidden Variable in Mathematics education? (pp. 13-37). Kluwer Academic Publishers, Netherlands.

Seppälä, R. (Ed.) (1994). Matematiikka – taitoa ajatella [Mathematics – a skill to think]. Suuntana oppimiskeskus 24. Opetushallitus. Helsinki.

Tobin, K. (1990). Changing metaphors and beliefs: a master switch for

teaching? Theory into Practice 29 (2), 122 - 127.

Välijärvi, J.& Sulkunen, S. (2012). PISA09. Kestääkö osaamisen pohja? Opetus- ja kulttuuriministeriön julkaisuja 2012:12 . Opetus- ja kulttuuriministeriö.

Wilson, M. & Cooney, T. (2002). Mathematics Teacher Change and Development. The role of beliefs. In G. C. Leder, E. Pehkonen, & G. Törner (eds.), Beliefs: a Hidden Variable in Mathematics Education? (pp. 127-147). Netherlands: Kluwer Academic Publishers.

Seppälä, R. (Ed.) (1994). Matematiikka – taitoa ajatella [Mathematics – a skill to think]. Suuntana oppimiskeskus 24. Opetushallitus. Helsinki.

Chilean and Finnish Teachers' Conceptions on Mathematics Teaching

Laia Saló i Nevado[1], Päivi Portaankorva-Koivisto[1], Erkki Pehkonen[1], Leonor Varas[2], Markku Hannula[1], Liisa Näveri[1]

University of Helsinki[1], Universidad de Chile[2]

Content

1 Background 44
2 Theoretical Framework 44
    2.1 Focus of the study 45
3 Implementation of the study 45
    3.1 The interviews 46
4 Findings 47
    4.1 The case of Nicholas 47
    4.2 The case of Amelia 49
    4.3 The case of Fiona 50
    4.4 The case of Dana 51
5 Conclusions 52
6 References 52

Abstract

This paper analyses the cases of two Chilean and two Finnish elementary teachers' reflections on their own professional development during a research project where they learned about and used open-ended problems to teach mathematics. The data indicates that during the project the teachers increased their pedagogical content knowledge, subject matter knowledge and motivational components. Teachers claim they give more room for pupils' ideas and rely on pupils' learning in pairs or in groups. Furthermore, they allege that also the weakest pupils seem to be involved with problem solving.

1 Background

This study is part of a larger research project in the field of mathematics education, financed by the Academy of Finland (project #135556) and Chilean CONICYT (project # AKA 09). The focal point of the project is teachers' professional development along with both teachers and pupils' development in mathematical thinking and understanding as well as pupils' performance when dealing with open-ended problems. The same problems are dealt with in both Chile and Finland. This paper is particularly focused on the teachers' own conceptions of their development and not on the results on pupils' development which are described in other papers (e.g. Laine & al. 2012, Varas & al. 2012).

The initiative for a joint research project came from Chile. The Academy of Finland and the Chilean CONICYT made an agreement for a research enterprise on education where mathematics education was a part of it.

2 Theoretical Framework

In mathematics education, problem solving is considered as a method to promote pupils' high-order thinking and understanding (e.g. Schoenfeld 1992). In the 1970's in different countries (cf. Nohda 1987), the methods of using open-ended problems were developed, in order to confront challenges of constructivism.

A problem is said to be open-ended, if it has an exactly stated starting point, but there are many possible goals where a solver might end, with equally correct methods (cf. Pehkonen 1995). Therefore, such problems do not have only one solution, but there might be many possible correct results, depending on the choices the solver has done.

When discussing the teaching of open-ended problems, the focus should be on teachers' pedagogical content knowledge and subject matter knowledge (cf. Shulman 1986). The use of open-ended problems challenges teachers to modify their roles in class. A teacher is no more a deliverer or transmitter of information, but a guide and facilitator for learning, and a planner of learning environments. Thus, teachers need to alternate and to improve their own conceptions of teaching and learning (cf. Pehkonen 2007). In order for teachers to be able to make such an adjustment, they need, for example, to learn to be sensitive to pupils' ideas and solution efforts, and to listen deeply to and try to pick up pupils' understanding (Stein & al. 2000). This all means a huge modification in teachers' pedagogical conceptions; the change might even be a radical one (cf. Merenluoto 2005).

A new idea for helping teachers to evolve is to accept that they are experts in developing new teaching solutions if only they are given the access to the newest results of theoretical studies (e.g. Wiliam 2002). This idea was empirically confirmed in the research paper by Roddick and Begthold (2004) where the authors found one of the critical change factors to be teachers' participation in the project from planning to implementation.

On the basis of previous research (e.g. Merenluoto 2001), the study at hand works with the hypothesis that when individuals encounter a phenomenon unknown to them, it leads to different cognitive effects and levels of learning. On the basis of the empirical research, another hypothesis sustains that teachers' prior understanding is quite resistant to the change (cf. Pehkonen 2007). Since the use of open-ended problems is difficult for most teachers, its learning takes time. However, there are other studies showing that such changes can also be very rapid (cf. Liljedahl 2010).

2.1 Focus of the study

The main question to be answered is the following: How do the teachers themselves perceive their conceptions about mathematics and teaching of mathematics have changed during the project?

3 Implementation of the study

The project is ongoing and will be implemented as a three-year (2010–13) follow-up study in the elementary school using quasi-experimental design. In the experimental group there are 10–20 elementary classes with their teachers from Santiago and Helsinki involved and committed to the study; and there is the same amount of classes in the control group in both countries. In these experimental classes, the teachers have implemented an open-ended problem once a month, from grade 3 to grade 5, in both countries. The researchers have followed the teachers' planning, implementation and reflection (self-evaluation) of their mathematics lessons, when the same open-ended problems are used in both countries.

In the following, there are three examples of open-ended problems used in the project:

1) Coloring a flag (grade 3): When using exactly three colors, plan as many different flags with three stripes as possible.

2) Snail-Elli (grade 4): Elli the snail climbs a wall very slowly. Some days the snail ascends 10 cm, some days she ascends 20 cm, some days she sleeps

and it does not move, and other days she is sound asleep and falls 10 cm. The wall has 100 cm of height. At the end of the tenth day Elli is at the half of the wall's height? What could have happened in the first 10 days? Show as many ways as possible.

3) Rectangles (grade 5): Find rectangles the perimeter of which is 16 cm. Find the area of each of them, and the rectangle with the biggest area.

Altogether there are seven problems per school year, i.e. about one problem lesson per month. The purpose of using such problems is to develop pupils' thinking skills and creativity. The main point is to let them see that there is not only one solution in a mathematical problem, but there might be many. Therefore, the problems dealt with are open, in the sense that they have many correct answers; very often the amount of correct answers is infinite. Additionally, the pupils are encouraged to find out the strategies of finding all solutions, if possible.

During the project, the problems were discussed with the teachers before and after the implementation in a joint meeting with the experimental teachers and the researchers' group in each country. In the meeting, they were talking about the experiences in the previous implemented problem and discussing about the next problem coming. In addition, in Finland there was a theoretical topic (around problem solving) presented to and discussed with the experimental teachers. In Chile, this was done in a two-week training seminar regarding open-ended problems previous to the beginning of the project for all the Chilean experimental teachers.

For teachers, the use of open problems means a new approach for teaching mathematics. They should modify their teaching habits, from much talking to more listening. And this might be for some teachers a huge conceptual change in their pedagogy, perhaps even a radical one (cf. Merenluoto 2005).

3.1 The interviews

For this paper, we analyze part of the research data that has been gathered through teacher interviews, classroom observations, videotaped discussions and the researchers' field notes. The teacher interviews took place during March-April 2012 in Finland and during November-December 2012 in Chile.

In this paper we analyze four of the cases, two teachers from each country, in order to reach a deeper level of analysis and more detailed reporting. The interviews of the Chilean teachers (Amelia and Nicholas) were selected from a sample of 10 interviews, and they were chosen because of their differences in a particular measurement regarding the quality of the introduction of a problem-solving task (arithmagon) that has been analyzed previously (see Varas & al.

2012). In Finland, only two teachers (Fiona and Dana) from 10 experimenting teachers volunteered to be interviewed. However, the measurement regarding the quality of the introduction of a problem solving task indicated that all four teachers were considerably different, which met our interest for answering the research question in different cases, therefore, providing a broader perspective (see Varas & al. 2012).

Interviews were conducted as semi-structured individual interviews that lasted between forty-five minutes and an hour and a half. The same questions were used in both countries, but because interviews were conducted with interviewees' native language two different researchers were used. Analysis of the interview data was carried out as a joint analysis in English.

The interviews were recorded. Each interview was divided into six different sections; in these the participants were asked about their background, the mathematical thinking of their pupils, the characteristics of their mathematics lessons, their involvement in the project, their expectations within the project and their professional development.

The observations and field notes were taken in Chile during November and December 2012, during visits to the different schools of the project and open discussions with the teachers and other school staff. In Finland the observations and field notes were from March, April and May 2012.

4 Findings

Here are the findings for the four teachers (Nicholas, Amelia, Fiona and Dana) based on the data collected through interviews, observations and field notes. The findings are gathered into the following table:

4.1 The case of Nicholas

Nicholas is a 49 year old Chilean male teacher. He claims that he gives freedom to his pupils to work, and listens and values pupils' ideas. In general, he believes that in his lessons he clearly becomes a mediator of pupils learning by being active and perceptive of the pupils' ideas and promoting the appearance of diversity in the pupils' production and responses.

Nicholas was obliged to join the project by the headmaster. At first, he had no expectations, and had a feeling there was nothing in it for him. He claims he felt tired of teaching. Nicholas considers he used to be a structured teacher in mathematics and having structured ideas about mathematics. However, he claims that during the project, he started to feel comfortable with mathematics. He feels that mathematics no longer has the same authoritative position and thus often there are no single right answers. On the contrary, he believes that people construct mathematics in a difficult way, and they should not be afraid of it. According to him, mathematics is everywhere and mathematics teaching is to enable the child to understand how things are formed. He says: *"God might have been a great mathematician, because wherever you look around... there is maths".*

Table 1: Findings regarding the Perceptions of four teachers: Nicolas (N), Amelia (A), Fiona (F) and Dana (the abbreviation used: OE = open-ended).

	BEFORE THE PROJECT	MIDDLE OF THE PROJECT
MATH	N: Math seemed to be an imperative A: Always liked math, but difficult and hard F: Math one subject among others, very easy D: Math structured, systematic and accurate	N: Enthusiastic about math A: Math difficult, but interesting F: Math connected to the children's world D: Perception has not changed, but deepened
OPEN-ENDED PROBLEM SOLVING	N: A problem needs to have only one solution A: Didn't know that OE problems existed F: Not familiar with the concept D: used problem-solving activities	N: OE problems are as valid as other type of math A: Sees the benefit in the use of OE problems F: OE problems are for all pupils D: Critical with some of the OE tasks
PUPILS CAPABILITIES	N: Had quite many weak pupils A: Pupils were mostly scared of math F: Wished the pupils to learn to connect problem solving with everyday's reality D: Sometimes pupils come up with unexpected solutions.	N: Some weak pupils had lost their fear to math A: Pupils seem more secure and like more math. F: Pupils think independently and not give up easily D: Pupils self-awareness has increased

	BEFORE THE PROJECT	MIDDLE OF THE PROJECT
TEACHING MATHS	N: Very structured. A: Liked to have control over the class and use concrete materials. F: Likes to lead when she teaches math, and follows the book D: Liked group work based lessons.	N: Not being an authority in the class A: Gives no room for pupils to improvise F: Has more patience to discuss and more independent from the book. D: Values more interaction with the pupils.

Nicholas declares that for him mathematics learning is based on the concrete, with games and through discovering. He insists on the importance of letting the pupils discover. After two years in the project he thinks his view of mathematics has changed. He claims that *"now math is crazy... and I owe it to the project"*. He complains that he does not like to plan the lessons and thus, the lessons result in what he calls a "salad of diverse activities". He describes himself as a mediator, and claims that the project has helped him understand how important it is for the teacher to allow pupils to express themselves.

For Nicholas a mathematical problem is a difficulty that a child has, and that the child can solve with mathematical attributes. With open-ended problems, Nicholas was surprised that the weakest pupils seemed to be motivated and involved. He considers it a challenge for the teacher to sustain the children's interest. According to him, by incorporating open-ended problems to his lessons, the result is a clear increase of participation and, pupils like mathematics more and develop their potentials better. He thinks that, in all, the project has affected his class positively and the pupils seem more open to mathematics. However, despite all the impact he claims that the use of open-ended problems has had in him, not much has changed in his lessons. The main change is that the pupils realize that he is fascinated with mathematics.

4.2 The case of Amelia

Amelia is a Chilean 47 year old female teacher. Amelia has low expectations of her pupils. Even though she is pessimistic, she is always ready to acknowledge the work of her pupils. Most of the time, the pupils' work is better than what she expects.

She got involved in the project because she thought it was interesting, and it was going to help the children. And as a result of it, she feels that there are children that are more secure and like mathematics more.

Amelia claims that mathematics is something that she has always liked, but it has been hard. She says that to teach mathematics is to give tools to the children to show the world of numbers, which is the practical and logical world where children live. Amelia remarks that mathematics is the base for everything to function well.

The use of open-ended problems has made Amelia realize that she speaks too much in her lessons, and she feels that she is mostly the main actor and thus, as a result, the pupils follow her way. She recognizes that she does not let them speak or think. She justifies that her intentions are to guide them through her experience.

Amelia believes that open-ended problems develop pupils' own abilities. She clarifies that with the traditional problems that are closed and structured, some pupils are not capable of solving them and getting engaged. Amelia claims that with open-ended problems, some of the weakest pupils are motivated and participate during the lesson.

In addition Amelia feels that, as a negative part, the open-ended problems do not connect with the contents of the curriculum, she is expected to teach. In her normal lessons she says she feels appalled to go back to fixed answers and structured ways to solve things.

4.3 The case of Fiona

Fiona is a Finnish 41 year old and claims that she uses partly teacher-led practices. She plans her lessons well and in her partly pupil-centered practices, her pupils work in pairs or groups. She explained that she has high expectations of her pupils and her classes are filled with work. She said that mathematics was an easy subject at school and she felt she did not need to put extra effort into learning it.

Fiona got the information of the project from her headmaster. She was inspired to get new ideas about teaching mathematics, professional development and using problem solving in her classes. Fiona claims that she used problem-solving tasks earlier in her mathematics teaching only now and then, and mostly as some extra material for her gifted pupils. However, within the project, she says she has learned a lot of how to use problem-solving tasks in her teaching and how to instruct her pupils.

According to Fiona, her perceptions of mathematics have changed during her working years. She claims that concreteness is of increased importance, and that she wants to tie mathematics into the lives of children. She insists that within the project she has learned to understand the value of guiding and asking relevant questions. She feels confident in herself: "I know I am on the right track!"

Fiona tells that her teaching of mathematics has also changed. She explains she has learned to separate the essential from irrelevant, and has now courage to vary her teaching methods. Fiona believes now that she can weed out material and achieve the same objectives with different methods also. She affirms to be more relaxed and she is confident on her professional skills.

There have been changes in Fiona's classroom practices, too. She mentions that now she understands that problem-solving tasks can promote mathematical thinking, and she can provide them to everyone, not just for the talented pupils. Fiona reveals that in her lessons, when pupils work in pairs, they help each other and their self-efficacy seems to have increased.

4.4 The case of Dana

Dana is a 49 year old Finnish teacher. She is an affective person, who appreciates creativity and spontaneity. Dana says that she does not plan her teaching that much beforehand but relies on pupils' ideas. She got the information of the project from her headmaster. Dana explained that she had used problem-solving tasks in her mathematics teaching now and then. But she adds that her own perceptions of mathematics have not changed much though during the project. She claims that she develops her pupils' mathematical thinking by questioning, and not giving ready-made answers. She tells that she gives nice feedback often and accepts many of her pupils' answers, even though they are perhaps not quite right. According to her, during her years of work, her mathematics teaching has changed. She tells about some kind of relaxation in her teaching and claims to live in the moment.

Dana explains that there have been changes in her classroom practices, too. Nowadays Dana reveals that she allows her pupils to work in pairs for the pupils to negotiate and consult each other. In her opinion some of the tasks used in this project were inspiring and during the project Dana insists she has learned to understand different areas of her pupils' interests.

5 Conclusions

As an answer to the research question "How do the teachers themselves perceive that their conceptions about mathematics and teaching of mathematics have changed during the project?" we have the following findings:

Firstly, the teachers claimed that the use of open-ended problems had an effect on their' conceptions about mathematics and teaching of mathematics since, for example, in the case of Dana, the subject knowledge increased and, in the case of Nicholas, the motivational component and his enthusiasm towards mathematics clearly increased.

Secondly, regarding the teacher's perceptions about pupils as mathematics learners, their allegations of transformation were evident as well. This is in line with the findings about pupils (Laine & al. 2012, Varas & al. 2012). For instance, Amelia noticed the growing interest in pupils and Fiona declared she was able to notice pupils' mathematical thinking. In addition, Nicholas and Amelia claimed to be surprised to see how also the weakest pupils can be involved with problem solving.

Lastly, all teachers declared that the use of open-ended problems seemed to have affected teachers' classroom practices up to certain extend. Nicholas, Fiona and Dana revealed the fact that they think they give more room for pupils' ideas, comments and arguments, and Fiona and Dana claimed that they rely on pupils' learning in pairs or in groups. Furthermore, Nicholas and Dana noticed how different tasks awaken different learning and interest. However, Amelia and Nicholas confessed that not much has changed in their daily lessons, and they blame this on the strictness of the system in Chile.

There seem to be no differences in the development of teachers' conceptions on mathematics and it's teaching in Chile and Finland. In both countries one may notice the teachers evolve in similar ways that seem to be due to the use of open-ended problems: teachers give more room for their pupils' ideas, comments and arguments, notice how different tasks awake different learning and interest, and that the weakest pupils can be involved with problem solving.

6 References

Laine, A., Näveri, L., Pehkonen, E., Ahtee, M., Heinilä, L. & Hannula, M.S. 2012. Third-graders' problem solving performance and teachers' actions. In: Proceedings of the ProMath meeting in Umeå (ed. T. Bergqvist), 69–81. University of Umeå.

Liljedahl, P. 2010. Noticing rapid and profound mathematics teacher change. Journal of Mathematics Teacher Education, 13(5), 411-423.

Merenluoto, K. 2001. Lukiolaisen reaaliluku. Lukualueen laajentaminen käsitteellisenä muutoksena matematiikassa. [Students' real numbers. Enlargements of number concept as a conceptual change in mathematics.] Dissertation. University of Turku, Ser. C 176.

Merenluoto, K. 2005. Discussion about conceptual change in mathematics. Nordic Studies in Mathematics Education 10 (2), 17–33.

Nohda, N. 1987. A study of 'open-approach method' in school mathematics. Tsukuba Journal of Educational Studies in Mathematics 4, 114–121.

Pehkonen, E. 1995. On pupils' reactions to the use of open-ended problems in mathematics. Nordic Studies in Mathematics Education 3 (4), 43–57.

Pehkonen, E. 2007. Über "teacher change" (Lehrerwandel) in der Mathematik. In: Mathematische Bildung - mathematische Leistung: Festschrift für Michael Neubrand zum 60. Geburtstag (Hrsg. A. Peter-Koop & A. Bikner-Ahsbahs), 349–360. Hildesheim: Franzbecker.

Roddick, C.D. & Begthold, T.A. 2004. Sixth grade mathematics teachers in transition: a case study. In: Proceedings of the PME-NA XXVI conference in Toronto (eds. D.E. McDougall & J.A. Ross), 1021–1028. Toronto: OISE / University of Toronto.

Schoenfeld, A.H. 1992. Learning to think mathematically: problem solving, metacognition, and sense making in mathematics. In: Handbook of research on mathematics learning and teaching (ed. D.A. Grouws), 334–370. New York: Macmillan.

Shulman, L.S. 1986. Those who understand: Knowledge growth in teaching. Educational Researcher 15 (2), 4–14.

Stein, M.K., Schwan Smith, M., Henningsen, M.A. & Silver, E.A. 2000. Implementing Standards-Based Mathematics Instruction: A Casebook for Professional Development, New York: Teachers College Press.

Varas, L., Näveri, L., Ahtee, M., Pehkonen, E., Fuentealba, A. & Martinez, S. 2012. Impact of different ways to introduce a problem solving task on pupils performance in Chile and Finland. Submitted as a presentation proposal for ICME-12 (TSG21) Seoul.

Wiliam, D. 2002. Linking research and practice: knowledge transfer or knowledge creation? In: Proceedings of PME-NA XXIV (eds. Mewborn, D.S., Sztajn, P., White, D.Y., Wiegel, H.G., Bryant, R.L. & Nooney, K.), Vol. 1, 51–69. Columbus (OH): ERIC.

Preschool Teachers' Conceptions about Mathematics

Lovisa Sumpter

School of Education and Humanities, Dalarna University, Sweden

Inhaltsverzeichnis

1 Introduction 56
2 Background 57
    2.1 Preschool curriculum 57
    2.2 Preschool teacher's conceptions and emotions 58
3 Method 59
4 Results and Discussion 60
    4.1 Static and dynamic emotional orientation 60
    4.2 What is mathematics at preschool level? 61
    4.3 What is a mathematical activity? 63
5 Summary 64
6 References 64

Abstract

This study looks at Swedish preschool teachers conceptions about mathematics and emotional directions towards mathematics. The results indicate that the preschool teachers are positive towards mathematics. When describing what mathematics is at preschool level, most teachers lists mathematical products such as mathematical concepts and procedures in arithmetic and geometry.

1 Introduction

Previous research has indicated the important role of preschool teachers' conceptions for the practice of teaching (Greenes, 2004; Lee & Ginsburg, 2007). What you express in some way guides or mirrors what you do. However, other research has shown that there might also be no or very little connection between beliefs and practice (Wilcox-Herzog, 2002). The disconnection in these cases can be a result of not having the ability to put the beliefs into practice, since the beliefs were more developed or advanced than the observed actions could reveal (Charlesworth, Hart, Burts, Thomasson & Mosley, 1993; White, Deal, & Deniz, 2004). In Sweden, a research study showed that 63 % of Swedish preschool teachers say that they have adequate knowledge in mathematics, whereas in language development (Swedish) 80% state they have sufficient knowledge (Sheridan, 2009). The limitations in the practice would then be in the education of the teachers – in the subject - more than the pedagogical ideas or the educational skills. If you then add the idea that "students learn what they have opportunity to learn" (Hiebert, 2003, p. 10), a statement in line with previous research showing that students that are not stimulated to practice processes do not develop corresponding competencies (e.g Bobis, et al, 2005; Bobis, Mulligan & Lowrie, 2008), a possible conclusion could be that mathematics at preschool level is influenced by teachers' conceptions and can be limited by their education resulting in different practices with different results. Benz (2012) ends her briefing on recent research dealing with this particular topic by saying that "we still don't know enough about the beliefs of people working in pre-school institutions especially concerning mathematics education." (Benz, 2012, p. 252). This study aims to add to the collective knowledge about preschool teachers' expressed mathematical world view by looking at Swedish preschool teachers' conceptions about mathematics and their emotional orientation towards the subject. Here we follow Thompson's (1992) definition of conceptions and see it as "conscious or subconscious beliefs, concepts, meanings, rules, mental images, and preferences" (Thompson, 1992, p. 132). This means that conceptions may have both affective and cognitive dimensions. There will be no attempt to try to separate different affective concepts from each other, for instance an attitude from a belief, meaning that they are treated in the research review and in the data analysis as 'conceptions'. The research questions posed are: (1) What is mathematics at preschool level according to Swedish preschool teachers?; and, (2) What emotional directions towards mathematics do they express?.

2 Background

2.1 Preschool curriculum

To understand the answers to the question 'What is mathematics?', we first need to know what mathematics is supposed to be according to the policy documents. Sweden got its first curriculum for preschools (children age 1-5) in 1998, but since then it has been revised. To illustrate the differences between the first curriculum (Lpfö 98) and the revised version (Lpfö 98/2010), we have made a table of the different goals to strive for (Table 1). Both curriculums starts with the phrase "The preschool should strive to ensure that each child...":

Table 1 Comparison of Swedish preschool curriculums

Lpfö 98	Lpfö 98/2010
develop their ability to discover and use mathematics in meaningful contexts	develop their understanding of space, shapes, location and direction, and the basic properties of sets, quantity, order and number concepts, also for measurement, time and change,
develop their understanding for basic properties in the concepts of number, measurement and shapes and their ability to orientate in space and time	develop their ability to use mathematics to investigate, reflect over and test different solutions to problems raised by themselves and others,
	develop their ability to distinguish, express, examine and use mathematical concepts and their interrelationships,
	develop their mathematical skill in putting forward and following reasoning,

Besides that the goals have increased from two to four, the mathematical concepts are in the revised version described in a more detailed way; e.g. instead of the single word number we have now sets, quantity, order and number as defined concepts. These are all examples of mathematical products. Also, in the revised version the emphasis is on the child's own construction of sense making, e.g. when developing the ability to distinguish, express, examine and use mathematical concepts and their interrelationships or developing their reasoning skills. They are examples of processes, or competencies. Mathematics at preschool level in Sweden has been given a new more emphasized role with this revised curriculum (School Agency, 2011). This follows a trend of 'schoolification' of early childhood education and care that we can see in several countries in Western Europe (Waller, Sandeseter, Ärlemalm-Hagsér & Maynard, 2010). Also, the policy documents stresses that it is the pedagogical professionals that are responsible for the development of children's joy to learn and to contribute to the realisation of the curriculum in everyday practice (School Agency, 2011), this including the goals for mathematics.

There is a growing body of research in early mathematics education based on the evidence that young children are more capable of developing mathematical concepts and processes than previously thought (Clements & Sarama, 2007). This has been further stressed by research that focuses on general mathematical processes such as problem solving, argumentation and justification (Perry & Dockett, 2008), modelling (English, 2012), and early algebraic reasoning (Warren & Cooper, 2008). Swedish research shows that five year old children can express mathematical competences such as reasoning and connecting mathematical entities with each other or representations of mathematical entities (Säfström, 2013). This would imply that young children's mathematics involves both products and processes. With this new curriculum, Swedish preschool teachers are in a new context facing different demands on the practice then before. Recent research shows that Swedish preschool teaching is indeed a profession in change, and that teachers create shared conceptions of how the new policies should be manifested in practice (Sheridan, Williams, Sandberg & Vuorinen, 2011). These conceptions are then the base for the children's learning.

2.2 Preschool teacher's conceptions and emotions

One of the first Swedish studies looking at preschool educators' conceptions was Doverborg (1987). She categorised the educators' responses into three themes: (1) mathematics is an activity in itself; (2) mathematics comes as a natural part in all situations; and, (3) mathematics is not an activity for preschool children. The most common reply from educators working in preschools (which purpose was to work toward schools) was that mathematics is an activity in itself. This could be a reflection of the goal that preschools had at that time. The most common response from the educators working in daycare (which offered child minding) was that mathematics is everywhere and it comes naturally in everything you do. Benz (2012) has studied German kindergarten professionals' attitudes towards mathematics in kindergarten. When asked about expected competences only a few replied that mathematics is a school activity and not a kindergarten activity. Most educators gave responses that could be interpreted as mathematical products. The most common adjectives used by the educators to describe their feelings towards mathematics in her study were useful (63%), important (59%), challenging (52 %), interesting (40%) and confusing (35 %) (Benz, 2012). Most of these adjectives could be considered positive, except for confusing and challenging were the first one is most likely to be negative and the latter could be either or. Research shows that there is a correlation between the attitudes towards mathematics and teacher's age, where younger teachers are more reluctant to the subject compared to older one's (Thiel, 2010). It could be a case of 'the more experience – the more you know' and with increased knowledge comes positive feelings.

Lee and Ginsburg (2007) have studied preschool teachers' beliefs about what they think is mathematics for 4 year olds. Teachers working in low and middle economic areas gave responses that emphasised teaching arithmetic and it has to be enjoyable. When providing examples of data of the first set of beliefs, it is mainly statements concerning arithmetic as concepts and procedures such as counting and one-to-one correspondence. This is similar result as Swedish preschool teachers' responses, but then with the addition of geometrical shapes (Sheridan, 2009).

3 Method

This present study is a pilot study for a larger research project aiming to look at Swedish preschool educators' conceptions and emotional directions towards mathematics. The data was collected in one cohort, in a small municipality in north of Sweden. The study was made through a written questionnaire handed out to 50 preschool teachers from five municipal preschools with children from age 1-5. The data consists or responses from 29 educators, all female. The response rate was then 58%. Of these 29, 24 were educated preschool teachers and five have vocational education from upper secondary school (where four of these have had some additional education, but not a teacher degree). They are all here treated as preschool teachers since they have the same responsibility for the children's educational activities. The questionnaire consists of 10 questions including four background questions. The questions that will be treated here are: [item 5] How would you describe your relationship (including emotions) towards mathematics?; [item 6] "What do you think is mathematics at preschool level?"; and, [item 7] "Do you work with mathematics in preschool? If yes, can you describe a typical situation when you work with mathematics?". The first listed question here aims to capture the preschool teachers' emotional direction towards mathematics, and the latter ones about their conceptions what mathematics is. This is a qualitative study. It works as a development of an analytic tool so that through thematic analysis we aim to identify themes that later can be used as factors (categories) for statistical analysis. A theme represents a patterned responses or meaning within the data set (Braun & Clarke, 2006) and can be identified in one or two primary ways in thematic analysis, inductive or deductive. Here we will use both an inductive and a deductive approach. We first use induction to see what categories data will generate in terms of the replies to what mathematics is at preschool level. As a second step, we will re-categories the responses in the categories provided by Doverborg (1987) and compare our results with her study.

4 Results and Discussion

4.1 Static and dynamic emotional orientation

To study preschool teachers' emotional orientation towards mathematics, we asked: "How would you describe your relationship (including emotions) towards mathematics?". All respondents answered the question with the following results (see Table 2):

Table 2 Preschool teachers' emotional direction towards mathematics

Emotional orientation	Number of respondents
Positive	22 (76%)
Negative	6 (21%)
Both pos/neg	1 (3%)

As we can see in Table 1, the majority of the preschool teachers (76%) state a positive relationship towards mathematics. These results are similar to Benz (2012). Five of the preschool teachers here describe their emotional orientation as a dynamic process where they previously have been negative towards mathematics, but now are feeling more positive:

> During my own schooling, mathematics wasn't interesting or an important subject. Especially during lower secondary school it just got more difficult and advanced and I lost the interest. When we now work with mathematics with the children in the preschool I realize that I have missed out, and I have got another attitude towards mathematics. I think it is fun and I see mathematics everywhere [P15]; I have always related the word mathematics to something negative. But, the more I work with it [the subject] and get inspiration from lecturers etc., the more fun it gets. [P21].

Both these teachers express an insight in the change of their attitude towards mathematics, and that the change is due to more experiences/ knowledge. This is in line with Thiel (2010) who reported that the attitude becomes more positive with time and experience.

4.2 What is mathematics at preschool level?

In order to look at preschool teachers conceptions about mathematics at preschool level, we asked "What do you think is mathematics at preschool level?". All respondents answered this question, and the responses were after inductive analysis divided into four categories were some replies fitted in to more than one category. The categories were mathematical products, "mathematics is everywhere", everyday activities, and the teacher's role.

Let us start with mathematical products. This was the most common reply; more than half of teachers (17 of 29, 58 %) mention mathematical products such as concepts and/or a procedure in their description of what mathematics is:

> This is something we use every day. We talk about the number of – how many, measure – how long? Find tall + short. Concepts about location on – under, behind – in front of etc. Look at the clock, showing what time [it is]. Shapes. Comparisons. Parts of e.g. an apple – ½ apple 1/8 apple ¼ apple [P29]; Make children curios about seeing shapes and their names. Make children curios about amounts and understanding for amounts. Make children curious about counting, measuring, compare sizes [P06].

The responses are mainly about arithmetic, just as Lee and Ginsburg (2007), with the addition of geometry, which is similar as the results from Sheridan (2007).

The second theme was "mathematics is everywhere"; 14 of the 29 preschool teachers (48%) gave a response that fitted into this category:

> A lot, Games, Dices, Music, Dance, Books, Physio etc. There is math in all situations. [P24]; Mathematics for me is everywhere. That could be e.g. concepts about location, amount, volume, equal different length, short. [P22].

Almost half of the teachers writes that "mathematics is everywhere", but as we can see from the replies we cannot say *why* mathematics is everywhere except from named situations or mathematical products.

The third theme was everyday activities. This theme could be described as a subset from the previous category, and 12 teachers (41%) gave an answer that was categorised into this theme:

Everyday situations when we set the table, when the children are building and constructing with different materials etc. [P19]; In everyday [situations] we are surrounded by math [P03].

The last theme was concerning the teacher's role. The 11 replies that fitted into this category were all a combination with one or several of the other three categories. Here are two replies that have both been sorted into both "mathematics is everywhere" and to "teacher's role":

Mathematics is everywhere at preschool level. You have to make it explicit.[P1]; Mathematics at preschool level shall be adjusted to the children's level, that could be everything from building with different types of blocks to sorting, comparing, categorisation. You can get mathematics in almost everything you do with the children. It is about consciously looking and listening for mathematics. [P21]

Summarising this part of the study, there seems to be an emphasis on mathematical products and in particular concepts from the mathematical areas arithmetic and geometry. Almost half of the preschool teachers say that mathematics is everywhere, and as a subcategory to this "everywhere", mathematics is something that exists in everyday life.

The next step of the analysis is to have a deductive approach and compare the responses with Doverborg's (1987) study (see Table 3):

Table 3 Comparison results with Doverborg's three categories.

Conceptions as themes	Doverborg (1987) - daycare	Doverborg (1987) - preschool	Present study
Mathematics is an activity in itself	15 (16%)	55 (80%)	1 (3%)
Mathematics comes as a natural part in all situations	61 (65%)	9 (13%)	26 (90%)
Mathematics is not an activity for preschool children	18 (19%)	5 (7%)	0 (0%)
Not classified	0 (0%)	0 (0%)	2 (7%)

Two of the responses could not be classified according to Doverborg's categories since they in their replies listed products (concepts and procedures) as a description what mathematics is at preschool level e.g.,

> Shapes, amount, ordering, sorting, sets many – few [P10]

Almost all of the respondents in this present study (26 of 29; 90 %) express a conception that fits into the category 'Mathematics comes as a natural part in all situations', e.g.,

> In principal, [mathematics is] everywhere. When we are at the preschool: when the children arrives, how many today, at breakfast; setting the tables for lunch, how many glasses etc. Sorting into categories how many belongs to what etc what belongs to what, how many etc. In the forest how far how many measuring steps sticks cones etc [P13]

As this reply illustrates, mathematics is according to this preschool teacher something that exists everywhere although it is not clear exactly what this 'everywhere' is beyond basic arithmetic (counting) and measuring. We need to find out more what this mathematics could be and in what way.

4.3 What is a mathematical activity?

To illustrate what a mathematical activity could be, we asked the preschool teachers to give an example of a typical situation. All 29 respondents say that they work with mathematics and gave several examples of typical situations. The situations named were e.g. eating (18 of 29 respondents: 62 %) and various assembly times (13 of 29 respondents: 45%). What is mathematics in these situations? In the replies from 24 of the preschool teachers we can get some information:

> Assembly, play in everyday situations! Outside! E.g. Singing time count your friends, say [what is] larger/smaller, who sits in front/back in between etc. [P11]; At fruit time, share the fruit in halves, quarters etc. At assemblies, singing. E.g. at physical education: use dice, say that with a one. Go 1 step with a two Jump 2 times 3: go backwards 3 steps [P18].

Again, the replies concern mathematical products such as counting and to giving name to objects. Even though most preschool teachers would emphasize the teacher's role in the situations, that they are the key to make things mathematical, it is still not clear from these results why eating or sharing a fruit in halves is mathematics. A result from this is that in the main study an addition of a clarifying question is added: what is the mathematics in the situation you just describe. Compared to the curriculums, both the first one from 1998 and the revised one (Lpfö 98/2010), we can see that the replies are more in line with the second goal from the first one: that the children should develop their understanding for basic properties in the concepts of number, measurement and shapes and their ability to orientate in space and time. It would be probable to think that the process of changing your mathematical world view when changed policy documents would take some time, following the results from Sheridan, Williams, Sandberg and Vuorinen (2011). From this present study we cannot say in what way preschool teachers' conceptions about mathematics encompass mathematical processes such as reasoning. A larger sample with an additional question may provide further information about that area.

5 Summary

According to the teachers responding in this study, mathematics at preschool level is products, mainly concepts from arithmetic and geometry, and the idea that "mathematics is everywhere", although it is not clear how and in what way mathematics is everywhere. When specifying situations, it is mainly different assembly times that were given as examples. For the majority of the respondents, mathematics is linked with positive emotions. Some teachers talk about a development from first being more negative to become more positive with time, as they gain experience and more knowledge. It is likely that knowing more helps to understand new policy documents and transform them in to practice. One implication of this study is that the teacher education should make sure that prospective preschool teachers (and maybe also in-service teachers) have the possibility to learn and explore what mathematical processes can be at preschool level. It seems like that is this stage that is the difficult part of mathematics education.

6 References

Benz, C. (2012). Attitudes of kindergarten educators about math. *Journal für Mathematik-Didaktik*, 33(2), 203-232.

Braun, V. & Clarke, V. (2006). Using thematic analysis in psychology. *Qualitative Research in Psychology*, *3*, 77-101.

Bobis, J, Clarke, B, Clarke, D, Thomas, G, Wright, R, Young-Loveridge, J. & Gould, P. (2005). Supporting teachers in the development of young children's mathematical thinking: Three large scale cases. *Mathematics Education Research Journal*, *16*(3), 27–57.

Bobis, J., Mulligan, J., Lowrie, T. (2008). *Mathematics for children: Challenging children to think mathematically*. (3rd ed.), Sydney: Pearson Education.

Charlesworth, R.C., Hart, C.H., Burts, D.C., Thomasson, R.H. & Mosley, J. (1993). Measuring the developmental appropriateness of kindergarten teachers' beliefs and practices. *Early Childhood Research Quarterly, 8*(3), 255–76.

Clements, D. H., & Sarama, J. (2007). Effects of a preschool mathematics curriculum: Summative research on the *Building Blocks* project. *Journal for Research in Mathematics Education, 38*, 136-163.

Doverborg, E. (1987). Mathematics in preschool? [Matematik i förskolan?] Publication 5. Gothenburg University, Pedagogiska institutionen.

English, L. D. (2012). Data modelling with first-grade students. *Educational Studies in Mathematics, 81*(1), 15-30.

Greenes, C. (2004). Ready to learn: Developing young children's mathematical powers. In J. Copley (Ed) *Mathematics in the Early Years*, (3rd ed.), 39–47. Reston, VA: NCTM.

Hiebert, J. (2003) What research says about the NCTM standards. In: Kilpatrick, J., Martin, G., and Schifter, D., (eds)., *A Research Companion to Principles and Standards for School Mathematics*, Reston, Va.: National Council of Teachers of Mathematics, 5–26.

Lee, S. L. & Ginsburg, H. P. (2007). What is appropriate mathematics education for four year-olds? Pre-kindergarten teachers' belief. *Journal of Early Childhood Research, 5*(1), 2–31.

Perry, B., & Dockett, S. (2008). Young children's access to powerful mathematical ideas. In L. D. English (Ed.), *Handbook of international research in mathematics education* (2nd ed) (pp. 75-108). New York: Routledge.

School Agency (2011). Curriculum for the Preschool Lpfö 98: Revised 2010. Stockholm: Skolverket.

Sheridan, S. (2009). Lärares och föräldrars syn på förskolan. [Teachers and parents' view on preschool] In *Barns tidiga lärande* [Children's early learning] S. Sheridan, I. Pramling Samuelsson & E. Johansson (Eds). (pp. 93-124). Gothenburg: Gothenburg University.

Sheridan, S., Williams, P., Sandberg, A. & Vuorinen, T. (2011). Preschool teaching in Sweden – a profession in change. *Educational Research* 53 (4), 415-437.

Säfström, I. (2013). Exercising Mathematical Competence: Practising Representation Theory and Representing Mathematical Practice., PhD thesis. Gothenburg University.

Thiel, O. (2010). Teachers' attitudes towards mathematics in early childhood education. *European and Early Childhood Education Research Journal, 18* (1), 105-115

Thompson, A.G. (1992). Teachers' beliefs and conceptions: a synthesis of the research. In D. Grouws (Ed), *Handbook of Research in Mathematics Teaching and Learning* (pp.127-146). New York: Macmillan Publishing Company.

Waller T., Sandeseter, E.B.H., Wyver, S. Ärlemalm-Hagsér, E., & Maynard, T. (2010). The dynamics of early childhood spaces: opportunities for outdoor play?, *European Early Childhood Education Research Journal, 18*(4), 437-443.

Warren, E., & Cooper, T. (2008). Generalising the pattern rule for visual growth patterns: Actions that support 8 year olds' thinking, *Educational Studies in Mathematics, 67*, 171-185.

White, C.S., Deal, D. & Deniz, C.B. (2004). Teachers' knowledge, beliefs, and practices and mathematical and analogical reasoning. In L.D. English (Ed) *Mathematical and analogical reasoning of young learners*, (pp. 127–152). New York: Lawrence Erlbaum Associates.

Wilcox-Herzog, A. (2002). Is there a link between teachers' beliefs and behaviors? *Early Education and Development 13* (1), 81–106.

Support or Restriction: Swedish Primary School Teachers' Views on Mathematics Curriculum Reform

Benita Berg, Kirsti Hemmi and Martin Karlberg
Mälardalen University

Content

1 Introduction 68
2 Data Gathering and Analysis 69
3 Results 71
    3.1   Analysis of Variance 72
4 Conclusion and Discussion 75
5 References 77
6 Appendix 79

Abstract

This paper reports the results of a quantitative study on primary teachers' (n=253) views on the introduction of steering documents and national examinations for Grade 3. While the majority of the teachers experience the reform as empowering, some teachers feel the new curriculum and the national examination restrict their teacher professionalism. We found differences in how teachers viewed the reform depending on whether they had graduated before or after the reform in 1994. The differing views can be connected to teachers' beliefs about teacher professionalism and the relation between teaching, learning and maturing. We discuss our findings also in the light of curriculum development during the last four decades.

Key words: teacher beliefs, curriculum reform, teacher professionalism, primary school, mathematics

1 Introduction

After a period of decentralization with weak classification and framing (see Hemmi & Berg, 2012; Bernstein, 1990), during which schools and primary teachers had a great deal of space to choose both the mathematics content to be dealt with and the rate of instruction in the classroom, the Swedish government has introduced new steering document (Lgr11) and national examinations that aim to support teachers by providing them with clearer guidelines concerning both content and knowledge requirements (Skolverket, 2012). Before the recent reform, only very general goals for Grades 5 and 9 in compulsory school were stated. The fulfilment of these goals was monitored for the first time during Grade 5, through national examinations testing a minimal level of pupils' achievement. Although the statements of goals and content and the introduction of the national examination at Grade 3 imply more steering than the previous system, the curriculum can still be seen as a relatively general framework, as it does not recommend teaching methods, textbooks, lessons plans or tests (cf. Hemmi, Lepik & Viholainen, 2013; Fowler & Poetter, 2004). Also, the knowledge demands for Grade 3 only state the minimum goals.

There is a mismatch between the intended curriculum given by policy makers and the implemented curriculum taught by the teacher in the classrooms. Most reforms in mathematics have been presented by a top down approach which ignores teachers' beliefs and pedagogical practices (Cuban, 1993). Several researchers have shown the importance of teachers' attitudes and beliefs in the success of reforming a curriculum. To understand what impact the reform could have on teaching and learning, it is important to understand how teachers experience and relate to the new steering documents (e.g. Handal & Herrington, 2003; Kleve, 2007; Remillard, 2005). Gjone (2001) suggests: "If a curriculum reform is to be successful (implemented and realized in practice), it must presumably reflect a general attitude of the teachers that changes are necessary or desirable" (author's translation) (p. 103). A teacher's interest in changing his/her way of teaching may affect the implementation of a curriculum (Fullan, 1991). An important issue for the individual teacher's actions concerns whether the teacher believes a change is possible in practice. Not only the teacher's experiences of what works and does not work in teaching, but also whether the change is consistent with the teacher's goals and interests, impact his/her actions (Hargreaves, 1994).

Kleve (2007) found three types of constraints that affect teachers' implementation of a curriculum. The first entails the teacher not believing in the reform but rather thinking that the way he/she has taught mathematics in the past is the best. This teacher will probably not implement the curriculum. The second constraint Kleve discusses is the teacher believing the reform is positive, but nevertheless not teaching according to the curriculum. Parents' expectations, students' demands, the work plan, and lack of time are limiting factors that exist among the teacher's beliefs about his/her teaching practice. The third type of constraint is the teacher believing in the reform and planning lessons with, for example, an investigative approach according to the curriculum, but keeping classroom activities fairly "traditional". Identifying constraining factors offers greater opportunities to improve teacher education (Kleve, 2007).

We have previously explored Swedish teachers' views on the recent curriculum reform through a number of qualitative studies (Hemmi & Berg, 2012), and found quite different attitudes and beliefs among the teachers. While most of the teachers were positive and regarded both the curriculum and the national examination as support for their teaching, some of the teachers felt the reform forced them to go against the natural maturation process of the child. Some teachers considered the level of curriculum goals and national examination as too low in order to find and support high-performing pupils. There were also teachers who wanted to have even more detailed curriculum guidelines. We connect these different views to different beliefs about teacher professionalism. In this paper we investigate the main result concerning the teachers' beliefs and attitudes, obtained from the qualitative studies, on a larger scale, and we focus on the following questions:

• How do primary teachers (Grades F-3) relate to the new curriculum and national examination?

• Are there differences in the views with respect to gender, age, how long one has worked as a teacher and the year of examination?

There is no consensus on the notion of beliefs in our research field (cf. Furinghetti & Pehkonen, 2003) and in many studies the lack of an explicit definition of beliefs and attitudes can make it hard to understand what is being investigated (Di Martino & Zan, 2011). We consider the teachers' beliefs about teaching, learning and teacher professionalism as a part of subjective knowledge consisting of conceptions, views and personal ideologies and we also recognise that the teachers' beliefs are more or less changing by influences of the culture of education (cf. Hannula, Lepik, Pipere & Tuohilampi, in press). Further, in this study we assume "the simple definition" of an attitude as "a general emotional disposition toward a certain subject" (Di Martino & Zan, 2001, p.18).

2 Data Gathering and Analysis

Using the results of the explorative qualitative studies entailing interviews and questionnaires, we formulated 40 statements about the new curriculum and the national examination using the teachers' authentic utterances as a source of inspiration (cf. Oppenheim, 1998). The teachers were asked to rate their agreement with the statements on a five-step scale. The questionnaire was distributed to all teachers teaching a preschool class or Grade 1-3 in an average Swedish municipality who participated in a meeting in August 2012. Hence, almost all teachers (n = 253) in the municipality responded to the questionnaire.

The teachers' ages ranged from 25 to 67 years (M = 49.5, SD = 10.0), and they had been working as teachers from one to 42 years (M = 15.9, SD = 11.5). A total of 95% of the teachers were female.

In this paper we focus on teachers' views, and use four item pools to investigate them. Four subscales were created with factor analysis (see tables 1-4, Appendix). The internal consistency (Cronbach's alpha) is presented after the abbreviated name of the subscale (in brackets). Also the factor loadings are presented.

The internal consistency was examined. Nunnally and Bernstein (1994, p 265) argue that internal consistency should be above 0.70 (Cronbach's alpha) for reliability to be regarded as satisfactory. The internal consistency was reasonably satisfactory for two of the subscales (Support, Restrict), and for the others it was rather low. For the subscale: The teacher considers the level of the curriculum goals and the national examination to be low (Goals Low), the internal consistency was low but all variables in the subscale correlated positively with at least one of the other variables. Although its internal consistency was low, the subscale was theoretically meaningful. Conceptually, the questions measured the same phenomenon, although this was not confirmed in the analysis of internal consistency. The subscale Clearness had low internal consistency, but was used on the same basis as Goals Low.

Teachers responded on a five-point Likert scale, ranging from "strongly disagree" to "strongly agree". Responses were coded from 1 to 5, and the scores for the questions in a subscale were summed to a score for that subscale. The score for the subscale then was divided by its number of variables, which means that the maximum points on all subscales is 5, no matter how many questions are included.

We also investigated differences depending on different background factors. We believed there could be different views of the reform if the teacher had worked only during the previous, very general steering document (teachers graduated 1994 or later: GA94) compared with the teachers who had experience about the earlier quite detailed curriculum (teachers graduated before 1994: GB94). Bulmer (1979) suggests that data is skewed if skewness is < -1 or > 1. Data is peaked or flat if kurtosis is < -1 or > 1. The skewness of the data and kurtosis of the data was mostly within the limit. Parametric statistical analyses (ANOVA) were used for the scales Restrict, Goals Low and Clearness. The distribution for the Support scale was skew and not within the limit for kurtosis (Bulmer, 1979), see Table 5. Accordingly to this the Mann-Whitney U test was used for the Support scale.

Table 1   Skewness and kurtosis for the dependent variables.

	Graduated < 1994 (N=123-158)		Graduated ≥ 1994 (N=78-85)	
Variable	Skewness	Kurtosis	Skewness	Kurtosis
Support	-1.16 S	.62	-1.66 S	3.03 K
Restrict	-.11	-.39	-.47	-.61
Goals low	-.33	-.17	-.50	-.01
Clearness	-.27	-.22	-.35	.10

Note: Skew and peaking values are marked with *S* and *K*.

We also investigated the possible differences in teachers' views depending on gender as well as the teachers' ages.

3 Results

In general, the teachers' views on the introduction of the steering document and national examination in Grade 3 were quite positive in both groups (the teachers who graduated before 1994 GB94 and the teachers who graduated in 1994 or after 1994 GA94), and they considered the guidelines quite clear (Table 6).

Table 2.   Means, standard deviations, median and quartile deviation of the dependent variables.

	Graduated < 1994 (N=123-158)		Graduated ≥ 1994 (N=78-85)	
Variable	M	SD	M	SD
Support	3.5 (*Mdn*)	1 (*qd*)	3.75 (*Mdn*)	1.13 (*qd*)
Restrict	2.93	0.86	3.22	0.93
Goals low	1.82	0.68	2.02	0.71
Clearness	3.06	0.88	3.27	0.78

Note: for the subscale Support the median (*Mdn*) and quartile deviation (*qd*) are presented. For the other dependent variables the mean and standard deviation are presented.

3.1 Analysis of Variance

Teachers who graduated in 1994 or later (GA94) were slightly more supportive of the reform of 2011 (see Figure 1); in particular, they supported the use of national tests. However, the difference was not statistically significant [z = 1.73, n.s.].

On the contrary (and this difference was statistically significant), teachers in the GA94 group criticized the reform for limiting teachers' professionalism to a statistically significantly higher degree than those in the GB94 group did; see Figure 2. [F=(1,201) = 5.03, p<.05].

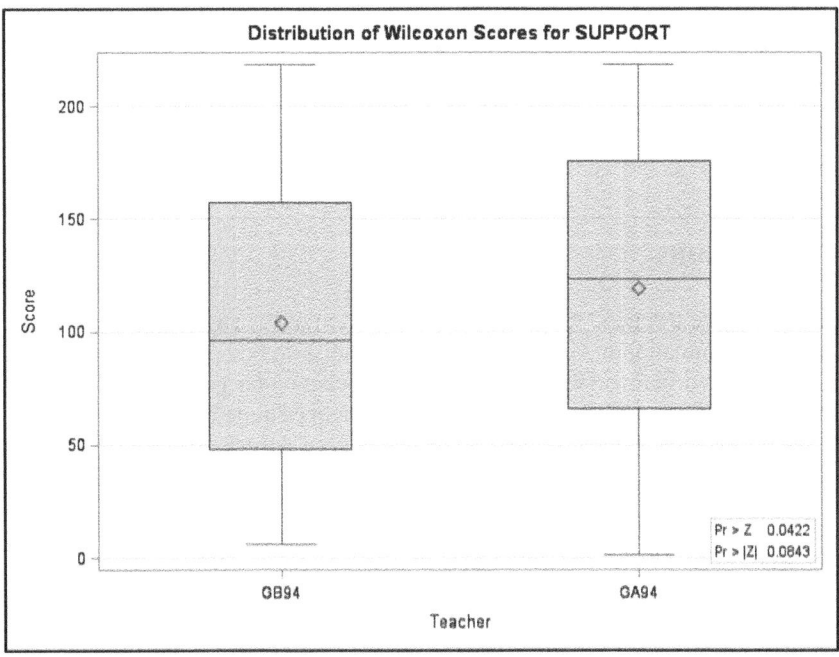

Figure 1   Support of the reform. Distribution of how positively teachers related to the reform in general and the national tests in particular. N=203

Swedish Primary School Teachers' Views on Mathematics Curriculum Reform 73

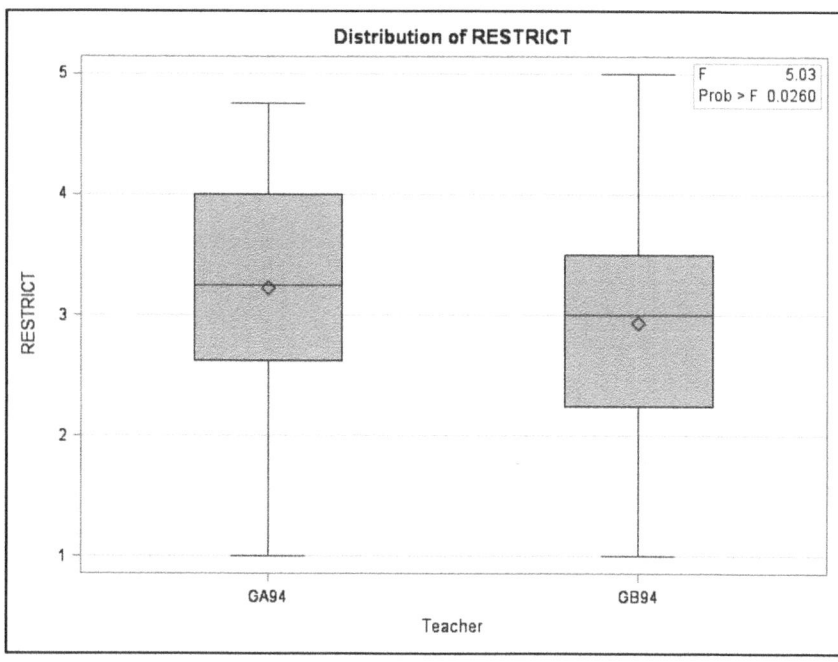

Figure 2   Restricting teacher professionalism. Distribution of experience of the reform as restricting teacher professionalism. N=212

These GA94 teachers, in comparison to those in the GB94 group, also considered the levels of the curriculum goals and the national examination to be low (see Figure 3), making it difficult to use the results from the national tests to identify and support high-performing pupils in need of extra stimulation. This difference between the groups was statistically significant [F=(1,208) = 4.56, p<.05].

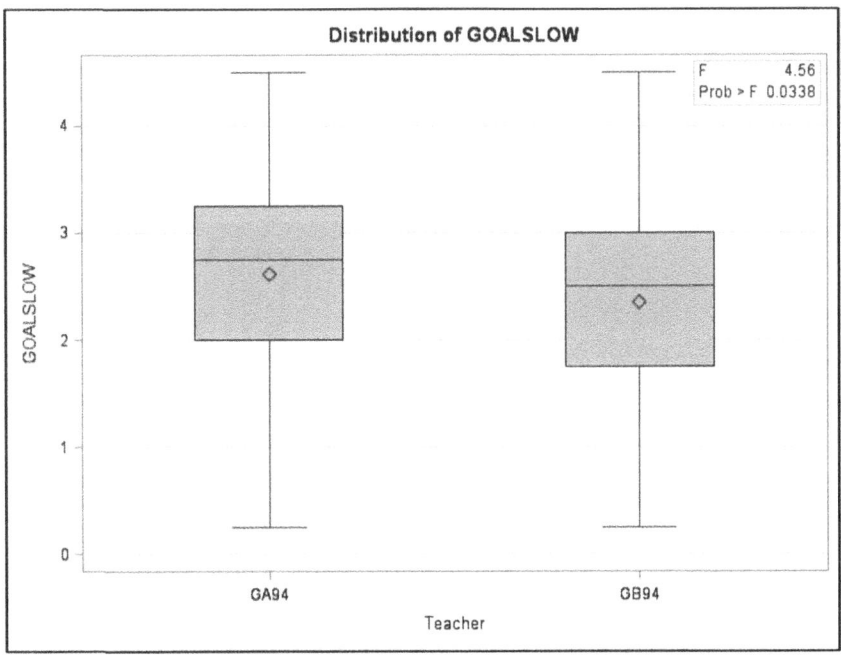

Figure 3   Low Goals. Distribution of the extent to which the teachers considered the level of the curriculum goals and the national examination to be low. N=219.

The teachers in the GA94 group thought that the steering documents gave specific support in planning and carrying out the teaching to a greater extent than those in the GB94 group did (see Figure 4). This difference was statistically significant $F=(1,221) = 4.61, p<.05$].

Complementary analyses showed no significant differences in the dependent variables between teachers who had worked less than ten years, ten to 25 years, and more than 25 years. Analyses with ANOVA and Mann-Whitney U test showed that the F-values for these analyses varied between .21 and .85. This indicates that a teacher's age does not influence his/her views on support for the reform, restrictions on their professionalism, the level of the curriculum goals, or the clearness of the curriculum guidelines. It rather seems to be that the time the teachers began teaching career is more important for their views on the reform of 2011. We also found no differences between the teachers' views with respect to gender.

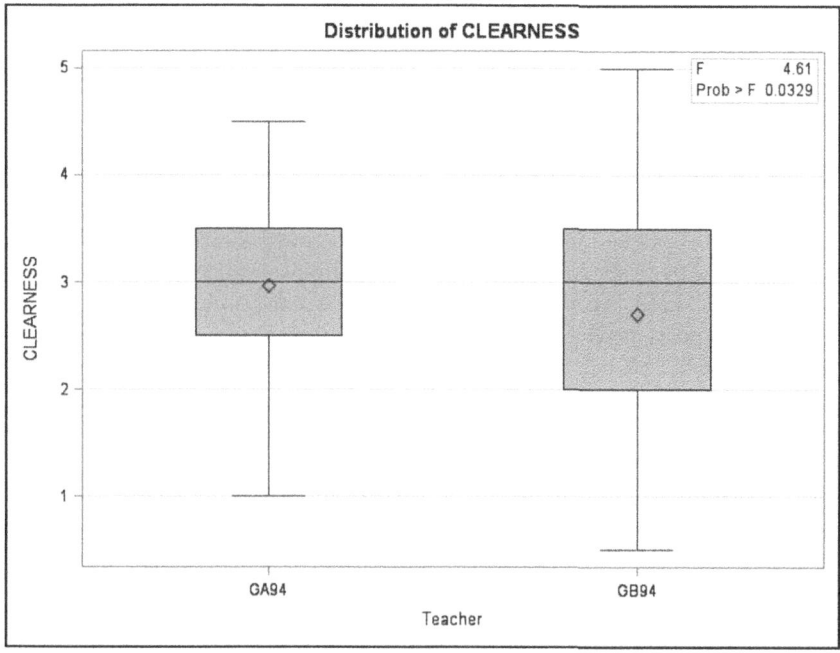

Figure 4   Clearness in the curriculum guidelines. Distribution of the extent to which the teachers viewed the curriculum guidelines as clear. N=223.

4   Conclusion and Discussion

Our study shows that most teachers in the municipality that we have investigated relate positively to the reform and, hence, the quantitative study confirms the results of our earlier qualitative explorations concerning this attitude (Hemmi & Berg, 2012). The most important background factor for different attitudes and beliefs seem to be the year of graduation. For example, the data analysis shows that the teachers who graduated in 1994 or later (GA94) were slightly more supportive of the reform, in particular the introduction of the national examination for Grade 3. These teachers (GA94) have worked only with very general national goals and guidelines, stated and measured for the first time for Grade 5, while the GB94 teachers have experienced also very detailed curriculum guidelines before the reform in 1994. It is possible that the GA94 teachers welcome the standards in form of the national examinations in order to relate their own goals to the national level while the GB94 teachers who started their teacher carrier during a period of more centralized guidelines might have continued using those guidelines as leading their work during the period of decentralization.

Against this background we find it interesting that the GA94 teachers felt that the reform restricted their teacher professionalism to statistically significantly higher degree than the teachers who graduated before the reform in 1994. It is possible that at the same time as the GA94 teachers find it helpful to compare their own goals to the national standards, they feel that the steering document with national examinations restrict their teacher professionalism, as they are used to work freely adapting the work on individual students' interests and developmental level. According to this view, it is impossible to state common mathematics goals proper for all students and therefore it is unnecessary to worry the pupils and their parents in vain. This view can be connected to teachers' beliefs about teaching, learning and the maturation of the child (see Vygotsky, 1978), something that may affect their views about what is possible to achieve in the teaching of mathematics.

The goals stated in the steering document for Grade 3 define the minimum level of student achievement, and the national examinations test only the fulfilment of these goals. A small group of teachers are concerned about this. We also found that this concern was more common among those who had graduated in 1994 or later, a fact we find difficult to explain.

Most teachers seem to be quite satisfied with the clearness of the guidelines. Again, we find an interesting difference between the GB94 and GA94 teachers. The teachers, who graduated in 1994 or later, feel that the steering documents offer them specific support in planning and carrying out the teaching to a greater extent than GB94 teachers feel. We think this can be connected to the GA94 teachers' earlier experiences about very general guidelines. The GB94 may still have in mind the older, quite specific guidelines and compared to those the new guidelines are quite general.

The teachers who advocate for clearer guidelines presumably also hold different beliefs about teacher professionalism, especially compared with those who believe that early goals are unnecessary and perhaps even damaging. If a teacher believes he/she knows the possible learning trajectories of the students and believes that it is impossible for all the students to learn the contents according to the national goals because of varying maturation of the children, he/she may not be willing to struggle to get all students to the decreed minimum level. According to earlier research teachers' positive attitudes towards reforms are important for the success of the reform (e.g. Kleve, 2007; Remillard, 2005; Gjone, 2001). Teachers' beliefs about a curriculum in relation to their own teaching can either facilitate or inhibit the realisation of the guidelines and ideas in new steering documents into classroom practices that reflect the reform (cf. Handal &

Herrington, 2003). Our results show that in general most of the teachers relate positively to the reform and hence, there should be good conditions for the reform to become successful. However, as Kleve (2007) points out, there can still be various external constraints that may have a negative impact on how teachers implement the curriculum, like parents' expectations or lack of time. Also, teachers can believe that the reform is needed and they try to adapt their teaching to the new demands. However, it is not always easy to introduce new contents in mathematics classrooms, even if the teacher would accept the importance of them. For example, problem solving is heavily stressed in the new Swedish curriculum and there are certainly a lot of teachers who are not accustomed to teach problem solving for children. That is why it is also important to focus on the character of the in-service teacher education where focus is on guidance about how to proceed in practice.

5 References

Bernstein, B. (1990). *The structuring of the pedagogic discourse: Class, codes and control.* London: Routledge.

Bulmer, M. G. (1979). *Principles of statistics.* New York: Dover Publications.

Cuban, L (1993). The lure of curricular reform and its pitiful history. *Phi Delta Kappan,* 75 (2), 182-185.

Di Martino, P., & Zan, R. (2001). The problematic relationship between beliefs and attitudes. In *Proceedings of the MAVI-X European Workshop, pp. 17-24. Kristianstad, Sweden.*

Di Martino, P., & Zan, R. (2011). Attitude towards mathematics: a bridge between beliefs and emotions. *ZDM Mathematics Education, 43(4),* 471-482.

Fowler, F. C., & Poetter, T. S. (2004). Framing French success in elementary mathematics: Policy, curriculum, and pedagogy. *Curriculum Inquiry, 34(3)*, 283–314 .

Fullan, M. (1991). *The New Meaning of Educational Change.* London: Teachers' College Press.

Furinghetti, F. & Pehkonen, E. (2003). Rethinking Characterizations of Beliefs. In G.C. Leder, E. Pehkonen, & G. Törner (Eds.), *Beliefs: A Hidden Variable in Mathematics Education.* (pp. 39-57). Secaucus, NJ: Kluwer Academic Publishers.

Gjone, G. (2001). Läroplaner och läroplansutveckling i matematik. In B. Grevholm (Ed.), *Matematikdidaktik – ett nordisk perspektiv.* Lund: Studentlitteratur.

Handal, B. & Herrington, A (2003). Mathematics teachers' beliefs and curriculum reform. *Mathematics Education Research Journal, 15*(1), 59-69

Hannula, M. S., & Lepik, M., Pipere, A., & Tuohilampi, L. (in press). Mathematics teachers' beliefs in Estonia, Latvia and Finland. *Proceedings of the Eighth Congress* of the European Society for Research in Mathematics Education. Antalya, Turkey. Feb 6th-10th, 2013.

Hargreaves, A. (1994). *Changing Teachers, Changing Times: Teachers Work and Culture in the Postmodern Age.* London: Cassell.

Hemmi, K., & Berg, B. (2012). Empowerment and control in primary mathematics reform – the Swedish case. In C. Bergsten, E. Jablonka & M. Raman (red.) *Evaluation and Comparison of Mathematical Achievement: Dimensions and Perspectives: Proceedings of MADIF 8.*

Hemmi, K., Lepik, M., & Viholainen, A. (2013). Analysing proof-related competences in Estonian, Finnish and Swedish mathematics curricula—towards a framework of developmental proof. *Journal of Curriculum Studies,* http://www.tandfonline.com/eprint/nzt5QqJAU2tqaVgya8eI/full.

Kleve, B. (2007). *Mathematics Teachers' Interpretation of the Curriculum Reform, L97, in Norway.* Unpublished Doctoral Dissertation. Agder University College, Kristiansand.

Nunnally, J. C., & Bernstein, I. H. (1994). *Psychometric theory.* New York: McGraw-Hill.

Oppenheim, A. N. (1998). *Questionnaire design, interviewing and attitude measurement.* London: Continuum.

Remillard, J. T. (2005). Examining Key Concepts in Research on Teachers' Use of Mathematics Curricula. *Review of Educational Research, 75*(2), 211-246.

Skolverket. (2012). *Ett år med en ny läroplan: Om Reformarbete, kunskapsbedömning och Skolverkets stöd. [One year with a new curriculum: About the reform work, assessment and the support of the National Agency.]* Stockholm: Elanders Sverige AB.

Vygotsky, L. (1978). *Mind in society: The development of higher psychological processes.* Cambridge, MA: Harvard University Press.

6 Appendix

Table 1. Variables in the subscale "The teacher relates positively to the reform in general and the national tests in particular" (Support, = .76) and the factor loadings of the variables.

Variable	Factor loadings
Results of national examinations will help me develop my teaching.	.59
National examinations are a great support for me as a teacher in my own assessment of students' knowledge.	.58
National examinations give me a good basis for individual development plans.	.45
National examinations give a clear picture of what knowledge the student has developed.	.54
National examinations concretize the goals of the curriculum.	.51
National examinations allow the teacher to diagnose what needs to be trained more.	.37
The goals for Grade 3 will help me get an idea of the student's development.	.38
The knowledge requirements help me focus on what is most important in my teaching.	.31

Table 2. Variables in the subscale "The teacher experiences the reform as restricting their professionalism as they think the students should develop at their own rate" (Restrict, = .65) and the factor loadings of the variables.

Variable	Factor loadings
National examinations needlessly worry students and parents.	.63
National examinations entail altogether too much work, you still know what students can do.	.48
National examinations make the students nervous.	.34
National examinations take a great deal of time to prepare, perform and process.	.31

Table 3. Variables in the subscale " The teacher considers the level of the curriculum goals and the national examination to be low" (Low Goals, $\alpha = .44$) and the factor loadings of the variables.

Variable	Factor loadings
National examinations do not help me identify students who need more stimulation.	.27
The goals are set too low for most students at the end of Grade 3.	.19
The tasks in national examinations are too easy to help me single out the students who can learn a lot more.	.31
It is difficult to set the bar higher because only the minimum level is described in the steering document.	.20

Table 4. Variables in the subscale "The teachers' views of the clearness of the curriculum guidelines" (Clearness, $\alpha = .47$) and the factor loadings of the variables.

Variable	Factor loadings
Lgr 11 gives me concrete support in how to plan and carry out my mathematics teaching.	.30
The steering document for Grade 3 should be more specific in order to guide me in my teaching. (reversed)	.30

Understanding Pre-Service Teachers' Belief Change during a Problem Solving Course

Carola Bernack-Schüler, Timo Leuders, Lars Holzäpfel
University of Education Freiburg (Germany)

Content

1 Introduction 82
2 Belief change: Theoretical framework and research 82
3 Research desiderata and goals of the study..83
4 Methodology 84
   4.1 Implemented course concept 84
   4.2 Interviews and Data analysis..85
5 Results and discussion..86
   5.1 Changing beliefs and stable beliefs 86
   5.2 Same belief - different reasoning 88
   5.3 Change of the belief structure 88
   5.4 Typology of belief change..90
6 Conclusion..91
7 References 92

Abstract

Different teacher programs and university courses in teacher education aim at changing the participants' beliefs towards a view of mathematics as a process and of mathematics learning as a constructivist act. They often include vivid experiences of problem solving and epistemic reflection. This paper describes a case study based on a pre-post interview design with pre-service teachers during such a course. One participant's belief system is analysed with respect to the beliefs before and after the course, the way of reasoning expressing these beliefs and the structure of the belief system. The results show amongst others that new beliefs evolve without the old ones being rejected. This leads to a more or less conscious ambiguity and to a conflict within the belief system. General results are briefly presented by describing a typology of belief change taking into account the whole sample of eight students.

1 Introduction

In school, pupils are not only learning mathematical content, they are also learning what mathematics is and how it is learned (Philipp, 2007). For that reason – and having in mind that many pre-service teachers have developed unhelpful beliefs during their school days – teacher educators aim at communicating a view on mathematics as a process and on mathematics learning as a constructive act (e.g. Schoenfeld, 1992). University courses which include for example vivid experiences of problem solving and epistemic reflection constitute an academic environment that can initiate and support belief change. It is of vital interest for designing such courses to better understand the processes of such a belief change. In this paper the belief change of pre-service teachers in a problem solving course is investigated to a deep level. The belief change in the course has already been corroborated inferentially by means of a questionnaire (Leuders, Holzäpfel, Bernack & Renkl., in prep.). However, the quantitative data used for that purpose cannot completely describe the complex characteristics of the change process in a belief system. Therefore, an interview study aimed to understand the belief change and to identify typologies. One case will be presented in a detailed way exemplifying the general results found in the other cases. The paper ends with an outlook on general results presenting a typology of belief change.

2 Belief change: Theoretical framework and research

Beliefs about mathematics and about mathematics teaching of prospective teachers develop primarily during their own school days where they are shaped by their experiences as students (Thompson, 1992). The change of belief systems occurs as a result of one's experience (ibid.). Green (1971) describes three dimensions of belief systems: First, beliefs have a quasi-logical relationship. Secondly, they can be central or peripheral which depends on their psychological strength. Finally they are held in clusters which can be more or less isolated from each other (ibid.). Green remarks that the disposition to change certain beliefs is related to the psychological strength of beliefs. Those that are located peripherally are easier to change than those located centrally (Green, 1971). Pajares (1992) argues in the same way though he differentiates between early and lately acquired beliefs. The question of why exactly it is so difficult for teachers to assimilate their schemes and to internalize new beliefs remains unanswered (Thompson, 1992). For that reason an important goal of research is to detect what happens cognitively when beliefs are changing and to identify those factors that promote belief change (Grigutsch, Raatz & Törner, 1998).

Several studies and teacher programs report belief changes, mostly towards mathematics as a process and towards a constructivist view of mathematics learning. These programs can essentially be characterized as follows: Working on open ended problems and/or unfamiliar mathematical content gives the participants the opportunity to experience and to reflect the role of a learner of mathematics (e.g. Chapman, 1999; DeBellis & Rosenstein, 2004; Liljedahl, Rolka, & Rösken, 2007; Lloyd & Frykholm, 2000). Some programs include participants' working with pupils and provide an opportunity to reflect on actual children's mathematical thinking (Ambrose, 2004; Chapman, 1999). Other programmes include the development and implementation of lesson plans (e.g. DeBellis & Rosenstein, 2004). Most of the programs aim at making the participants aware of their beliefs as well as providing emotion-packed experiences of doing mathematics. In almost all programs reflection is an important component. However, most of the studies focus on beliefs about teaching and learning mathematics or they do not clearly separate between beliefs about mathematics and those about teaching and learning. While all studies drew on retrospective reflection, only some studies refer to pre and post data to describe the belief change (Ambrose 2004; Chapman, 1999; Liljedahl, Rolka, & Rösken, 2007). A belief change of participants is reported throughout all these programs – few studies also use inferential analysis of quantitative data (e.g. Roscoe & Sriraman, 2011; Leuders et al., in prep.). Ambrose (2004) describes the change in the participants' belief system concluding "that prospective teachers do not let go of old beliefs while they are forming new ones" (ibid., p.117). With this caveat in mind, it seems reasonable to further investigate the quality of belief change.

3 Research desiderata and goals of the study

In the teacher education course reported in this paper, the questionnaire data demonstrated a belief change with respect to several dimensions (Leuders et al., in prep.). However these results neither explain the reasons for the belief change nor do they provide insight into the complex change of a belief system. Different beliefs concerning mathematics itself and the teaching and learning of mathematics interact in a complex way and thus can only be roughly captured by means of questionnaires. Stahl (2011) doubts the validity of capturing beliefs by questionnaire ratings because the same judgments might be given e.g. at different age levels for different reasons. Stahl pointed out that "the argumentation structures and the cognitive elements that different learners activate to reach the judgment might be greatly different from each other" (Stahl, 2011, p.42). Cooney (1998) proposed four types of belief change based on organizing beliefs in isolated clusters to creating a coherent belief system through reflection. However, he based his research on the case studies of four individual teachers, so that his findings must be judged critically. Furthermore, when belief change is reported from a retrospective view there may be bias in describing the change process. In order to overcome this bias the study reported here aimed to better understand the belief change and the individual belief system in a longitudinal design in relation with problem solving courses by means of interviews. A second goal consisted of identifying typologies of belief change which occur within a problem solving course. These objectives lead to the following research question: In which way does active problem solving and reflection of one's own problem solving processes influence beliefs about mathematics and beliefs about teaching and learning mathematics? We try to answer this overarching question by investigating the following sub-questions: Which beliefs change and which ones stay the same? How do the arguments used to express beliefs change? How does the structure of the belief system change? Which types of change can be found?

4 Methodology

4.1 Implemented course concept

The course concept draws on ideas from teacher programs like the ones reported above (e.g. DeBellis & Rosenstein, 2004). The research project is closely linked with a university course for future primary level teachers titled 'Mathematical Thinking'. The students work individually on open ended problems which lead to activities that involve exploring arithmetic or geometric phenomena, proposing and testing hypotheses, thereby generating systematic knowledge about mathematical structures. As one example of the problems used in the course the problem 'Step-Numbers' is presented in Figure 1:

Problem 3: Which numbers can you write as the sum of consecutive natural numbers (e.g. 12 = 3+4+5)? Can you tell which numbers can be written in which different ways? When you have worked on the problem to your satisfaction, ask some questions, e.g. "What happens if...?" or vary the problem.

Figure 1   Example of the problems

The participants (N=78, mainly in their second year of three of teacher education studies) received neither heuristic support nor any other kind of feedback from the instructor, so they could experience for themselves being independent and self-regulated problem solvers. They were required to keep records of the problem solving process including all ideas, emotions and reflections. At the beginning and at the end of the course the participants reflected on their notion of 'mathematical thinking' and on their beliefs by means of concept maps and written reflections. The course took place in an experimental intervention design involving a quantitative pre-, post-study on belief change (Leuders et al., in prep.). Moreover qualitative data on the problem solving processes and the participants' reflection was gathered. This paper focuses exclusively on the interview data.

4.2   Interviews and Data analysis

The interview study was based on a convenience sample; eight students from the course volunteered. All of them were studying to become primary or lower secondary school teachers. The interviews were conducted in a pre-post-design as guided interviews to avoid gathering only potentially distorted retrospective view. Because the goal was to understand the individuals' belief systems and their change, in each interview the same open questions about the nature of mathematics, mathematics at school and at university etc. were asked. Each interview took 45 to 70 minutes.

The method of analysis follows the Qualitative Content Analysis (cf. Mayring, 2000). The process of analysis mainly consisted of loops of summarizing and structuring according to areas and belief-categories which were identified both inductively and deductively. In the first step, the interview was divided into coherent text sections which then were classified by seven areas: 'Nature of mathematics', 'Mathematical tasks and problems', 'Mathematical activities', 'Context of own school days', 'Context of own studies', 'Mathematics learning and teaching' and 'Mathematical self-concept'. The selection of these areas for the interview was based on the dimensions and aspects of belief structures reported in literature (e.g. Op't Eynde, de Corte & Verschaffel, 2002). In the next step all sections of every single area were summarized. These thematic summaries were then categorized according to a set of belief categories. On the one hand, established belief categories (Baumert et al., 2006) were used if appropriate. On the other hand categories were inductively introduced to describe the individual belief system when new aspects arose. On this basis the following comparisons and interpretations according to the research questions could be carried out and will be presented in this paper: A comparison of pre- and post- for each student (exemplified for one case) as well as finding typologies of belief change (presented for all cases). Additionally, changes in the way of reasoning can be found and the change in the overall belief structure of the individual student can be analysed.

5 Results and discussion

5.1 Changing beliefs and stable beliefs

The case of Maria (pseudonym) will be presented here in a detailed way, so that the power of the method of analysis for further cases becomes apparent. Maria is 22 years old; she was studying to become a lower secondary school teacher and she was in her third year of the teacher education program. Her high school results in mathematics corresponded to an American grad A. During the first interview the belief of 'Mathematics as a toolbox' strongly appears within different areas. In Table 1 a cross indicates in which area this belief is expressed in the first (pre) and the second (post) interview. In the first interview 'Mathematics as a toolbox' appears e.g. within the area 'Tasks and problems'.

Table 1: The beliefs 'Mathematics as a toolbox' and 'Usefulness of mathematics' found within thematic summaries

Beliefs / Areas	M. as a toolbox		Usefulness of m.	
	pre	post	pre	post
Tasks and problems	x	x		
Context of her own school days			x	
Mathematics' learning and teaching	x	x	x	x
Mathematical activities	x	x	x	x
Nature of mathematics	x	x	x	x

Maria preferred to solve tasks by a known algorithm and tasks that focus on computation. She explained a typical task in mathematics as follows (a selection of different tasks was presented to her):

    Interv.: And why are they [these tasks] typical?

    Maria: It's clearly said what to do and referring to teaching, the student knows immediately what to do without having to resolve a puzzle – like in the other task.

    Interv.: And when not referring to teaching?

    Maria: Hm, generally spoken this task includes what is given and what is asked for, it is clearly understandable and yes, it is universally valid.

This belief still appears as stable during the second interview after the course within the same areas, but sometimes with little shifts and constraints. Asked about typical tasks in mathematics she explains for example:

> Maria: For this, I feel ambivalent. On the one hand, I think, tasks like [these] ones are typical because they specify what to do. On the other hand, I believe there is a big part in mathematics that tries to appreciate individual solution processes. I can't tell about the relation between these, that's difficult. I think, the own school days rather shape the idea what mathematics is. And perhaps I still have internalized that there has to be a formula.

Generally, the interviews reveal that old beliefs are mostly not rejected even if new ones are appearing.

Another strong belief during Maria's first interview was the 'Usefulness of mathematics' (cf. table 1, col. 3). She evaluated her former mathematics lessons as positive when she could transfer the content, e.g. geometry, to her everyday life. Her future pupils, she believed, should experience mathematics as something that is useful. In the second interview she focussed less on that belief. This is in line with the other interviewees where this belief remains or is less strong.

In the second interview Maria referred to an aspect of mathematics that did not occur in the first interview. Her statements can be categorized as 'Explorative activities', 'Individuality when doing mathematics', 'Different approaches' and 'Problem solving'. These categories can be summarized as a process oriented view on mathematics. They can be found in the areas 'Tasks and problems', 'Mathematics' learning and teaching','Mathematical activities' and 'Nature of mathematics'. In the quotations above from the second interview one can recognize the categories 'Different approaches' and 'Individuality when doing mathematics', when she is talking about individual solution processes. These beliefs are closely related to her experiences in the course. As it can be seen from the numerous areas within which she mentioned it, she is integrating them in her belief system. Asked what she is doing when she is practicing mathematics she refers to the course:

> Maria: One tries somehow to find connections between what is given. I realized that I focus on ALWAYS finding a formula. Even when you don't need one. I think, that is due to my school days where you are shaped to believe that for every task there is an ideal solution and a formula to be applied. It was a great experience that it can work differently, WITHOUT a formula. [...]
>
> Interv.: How could you describe this experience – of not using formulas?
>
> Maria: At the beginning, I didn't have an exact idea about the result. In the course, I often started to create examples, then I discovered regularities which I investigated and then I found a solution. And this went on like pearls on a string. In the end you find something great and in between again and again little things.

5.2 Same belief - different reasoning

Another belief-cluster can be found in Marias beliefs about teaching and learning. In the first interview she tells about 'Constructivist learning through discovery' with regard to the areas 'Context of her own school days', 'Context of university studies' and 'Mathematics' learning and teaching'. Amongst other things it seemed important to her that the content is discussed and discovered by the pupils. This belief was found in the same areas in the second interview but she was answering in a more precise and extensive way, using fewer buzzwords then in the first interview. She described the change to this belief as follows (excerpt from the second interview):

> Maria: Basically, I knew that pupils should work on tasks by themselves. But I never understood how that should work because I hadn't understood by myself how to tackle it when I'm uncertain about it. Now I realized that you don't need to bother and that this comes up intentionally.

After the course, almost all interviewees are expressing their beliefs in a more detailed and reflected way, especially related to the areas 'Mathematical activities', 'Nature of mathematics' and 'Mathematical teaching and learning'.

5.3 Change of the belief structure

Regarding the structure of Maria's belief system a certain ambiguity can be recognized. Referring to Green (1971) 'Mathematics as a toolbox' appears to be a central belief with more psychological strength then other beliefs about mathematics. Beliefs that can be summarized under the aspect mathematics a s process (e.g. 'Explorative activities' and 'Individuality when doing mathematics') are still new beliefs. Reading the second interview's excerpts above and considering further passages, an ambiguity between those two beliefs appears. Maria seemed to be consciously referring to this ambiguity that provokes an inner conflict. She recognized both apart from each other but she could not resolve the conflict or is restricting their application to certain aspects of mathematics.

> Maria: There are different areas in mathematics, one is for example problem solving where the solution procedure is rather open-ended. Everybody can develop its own approach. And for example in geometry, everything is more limited by formulas.

Regarding the seven cases one can find indications of different belief structures with regard to the aspect of ambiguity as in Maria's case – with varying sophistication and consciousness. These first results merit deeper analysis since these ambiguities can play an important role for example in the students' future teaching practice.

To summarize the results, Figure 1 tries to give an overview of Maria's central beliefs (darker boxes) and some local respectively less dominant beliefs (lighter boxes). The Figure additionally shows further belief categories and emotions towards mathematics that were found in the interview data and that could not be part of the in-depth description above. Figure 2 tries to describe her belief system after the course.

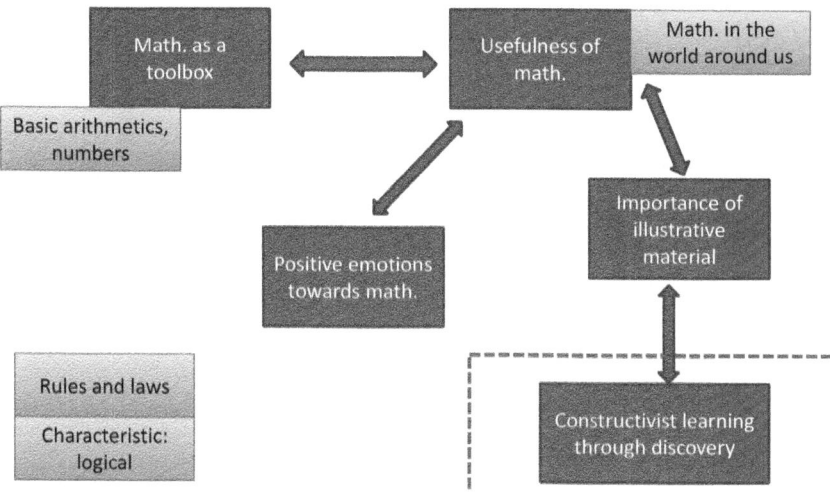

Figure 2   Maria's belief system before the course

The belief 'Explorative activities / individuality when doing mathematics' could only be found in the statements of the second interview. The new beliefs around 'Explorative activities' are related to 'Constructivist Learning' which is indicated by an arrow. Since it was the explicit goal of the course to strengthen such a view of mathematics it can be regarded as successful. However, the evolution of this new belief did not lead her to completely reject her old beliefs. The ambiguity between these new beliefs and the old one 'Mathematics as a toolbox' is highlighted by a flash between them in Figure 3.

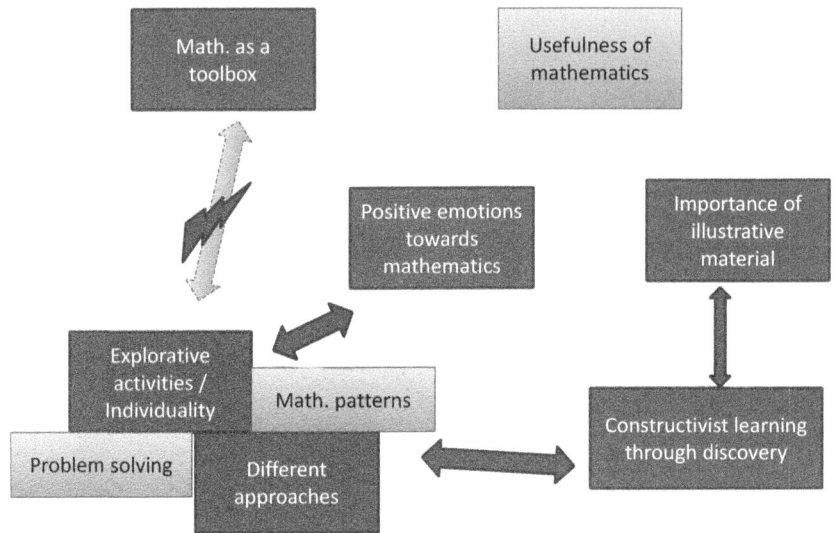

Figure 3   Maria's belief system after the course

5.4 Typology of belief change

In this article an in-depth analysis of a single case is reported. Further seven cases of participants were analysed in the same way revealing similar results with different individual emphasis. In order to provide more elaborated results we identified types describing the change process (Bernack-Schüler, in prep.). As criteria we chose the way beliefs connected to Mathematics as a process and those about 'Mathematics as toolbox' changed because they appeared of particular importance in the data and regarding the goals of the course. The combination of these two criteria revealed three types of belief change. Furthermore those types were characterized by means of the way of reasoning pre and post and the structure of the belief system after the course. The characteristics of each type are represented in Table 2. Maria belongs to Type 1. Type 2 is characterized by insisting on mathematics as a toolbox without developing process oriented beliefs that play a central role in their belief system. Type 3 stands out due to the aspect 'mathematics as a process' which was already important to them before the course and then confirmed during the course. Mathematics as a toolbox seemed not to be of particular importance in both interviews.

Table 2: A typology of belief change

	Type 1	Type 2	Type 3
Math. as a toolbox: Way of change and importance pre-post	Beliefs stay constantly important / Strengthening of existing beliefs	Beliefs stay constantly important / Strengthening of existing beliefs	Weak beliefs – no particular importance pre and post in the belief system
Mathematics as a process: Way of change and importance pre-post:	Newly evolved after the course	Weak beliefs – no particular importance pre and post in the belief system	Strengthening/ confirmation of existing beliefs
Way of reasoning	Vague or uncertain answers in the first interview	Vague or uncertain answers in the first interview	„Mathematics are everything and everywhere"
Belief system (post)	In some cases ambiguity found		integrated, reflected

6 Conclusion

This paper describes a single case as an example from an interview study of belief change during a university course in problem solving, together with insights from the general results. The results show that during a belief change new beliefs evolve or some beliefs are strengthened without the older ones being rejected – as it was already indicated by Ambrose (2004). This process may lead in certain cases to a belief structure that can be characterised by a more or less conscious ambiguity or an inner conflict. This finding also has to be considered when teacher educators implement such a course concept and it stresses the importance of reflection that Cooney (1998) describes in his paper. Furthermore, the example underpins the criticism by Stahl (2011): Belief structures may be more complex than judgements elicited in questionnaires reveal. Furthermore the same belief can be expressed with different degrees of reflection and reasoning. Interview studies are limited to the distinct questions posed and the answers elicited. There are possibly hidden beliefs that could not be made explicit by the participants. The case of Maria exemplifies type one of the belief change through such a problem solving course. The presented typology completes the results under a general viewpoint by taking into account all interviewees. More elaborated descriptions of the general results and case studies exemplifying type 2 and 3 can be found in Bernack-Schüler (in prep.).

7 References

Ambrose, R. (2004). Initiating change in prospective elementary school teachers' orientations to mathematics teaching by building on beliefs. Journal of Mathematics Teacher Education, 7(2), 91–119.

Baumert, J., Blum, w., Neubrand, M., Klusmann, U., Brunner, M., Jordan, A., ··· (2006). Professionswissen von Lehrkräften, kognitiv aktivierender Mathematikunterricht und die Entwicklung von mathematischer Kompetenz (COACTIV): Dokumentation der Erhebungsinstrumente. Berlin.

Bernack-Schüler, C. (in prep.): Entwicklung von Mathematikbildern bei Studierenden.

Chapman, O. (1999). Inservice Teacher Development in Problem Solving. Journal of Mathematics Teacher Education, 2, 121–142.

Cooney, T. J., Shealy, B. E., & Arvold, B. (1998). Conceptualizing Belief Structures of preservice Secondary Mathematics Teachers. Journal for Research in Mathematics Education, 29(3), 306–333.

DeBellis, V. A., & Rosenstein Joseph G. (2004). Discrete Mathematics in Primary and Secondary Schools in the United States. ZDM: The International Journal on Mathematics Education, 36(2), 46–55.

Green, T. F. (1971). The activities of teaching. McGraw-Hill series in education Foundations in education. New York: McGraw-Hill.

Grigutsch, S., Raatz, U., & Törner, G. (1998). Einstellungen gegenüber Mathematik bei Mathematiklehrern. Journal für Mathematik-Didaktik, 19(1), 3–45.

Leuders, T., Holzäpfel, L., Bernack-Schüler, C., Renkl, A. (in preparation). Getting the right picture by doing mathematics – Changing beliefs through individual mathematical inquiry.

Liljedahl, P., Rolka, K., & Rösken, B. (2007). Affecting Affect: The Reeducation of Preservice Teachers' Beliefs about Mathematics and Mathematics Learning and Teaching. In W. G. Martin, M. E. Strutchens, & P. C. Elliott (Eds.), The Learning of Mathematics. (pp. 319–330).

Lloyd, G. M., & Frykholm, J. A. (2000). How innovative Middle School Mathematics can change Prospective Elementary Teachers' Conceptions. Education, 120(3), 486–580.

Mayring, P. (2000). Qualitative content analysis. Forum Qualitative Sozialforschung/ Forum: Qualitative Social Research, 1(2), Art. 20, from urn:nbn:de:0114-fqs0002204.

Pajares, F. M. (1992). Teachers' Beliefs and Educational Research: Cleaning Up a Messy Construct. Review of Educational Research, 62(3), 307–332.

Philipp, R. A. (2007). Mathematics Teachers' Beliefs and Affect. In F. K. Lester (Ed.), Second handbook of research on mathematics teaching and learning (pp. 257–314). Charlotte, NC: Information Age Publ.

Op't Eynde, P., Corte, E. de, & Verschaffel, L. (2002). Framing students' mathematics-related beliefs. In G. C. Leder, E. Pehkonen, & G. Törner (Eds.), Beliefs. A hidden variable in mathematics education? (pp. 13 – 37). Dordrecht: Kluwer Acad. Publ.

Roscoe, M., & Sriraman, B. (2011). A quantitative study of the effects of informal mathematics activities on the beliefs of preservice elementary school teachers. ZDM: The International Journal on Mathematics Education, (43), 601–615.

Schoenfeld, A. H. (1992). Learning to think mathematically: Problem solving, metacognition, and sense making in mathematics. In D. A. Grouws (Ed.), Handbook of Research on Mathematics Teaching and Learning (pp. 334–370). New York: Macmillan.

Stahl, E. (2011). The Generative Nature of Epistemological Judgments. In J. Elen, E. Stahl, R. Bromme, & G. Clarebout (Eds.), Links between beliefs and cognitive flexibility. Lessons learned (pp. 37–60). New York: Springer.

Thompson, A. (1992). Teachers' Beliefs and Conceptions: A Synthesis of the Research. In D. A. Grouws (Ed.), Handbook of Research on Mathematics Teaching and Learning (pp. 127–146). New York: Macmillan.

Teachers' Beliefs Systems Referring to the Teaching and Learning of Arithmetic

Katinka Bräunling, Andreas Eichler
University of Education Freiburg, Germany

Content

1 Introduction 96
2 Theoretical framework 97
3 Method 98
4 Identifying central beliefs 99
5 Explaining peripheral beliefs 102
6 Characterisation of teachers' belief systems 103
7 Discussion 105
8 References 106

Abstract

In this paper, we focus on belief systems of six teachers of primary and secondary schools who just started their teacher trainees referring to the teaching and learning of arithmetic. Firstly, we discuss the theoretical framework of our research and outline the method. Afterwards we discuss findings of our research in three separate sections. We discuss the identification of central beliefs referring to one teacher. Further we derive findings referring to peripheral beliefs to the same teacher. Finally, we discuss types of belief systems towards teaching arithmetic. We conclude the paper with a brief summary and suggestions for further research.

1 Introduction

Teachers' decide, *what* mathematical content they bring to the classroom; they have reasons, *why* they select specific content and – except for ad-hoc decision when interacting with students – they decide *how* they teach specific content, i.e. they individually define their way of teaching (cf. Calderhead, 1996). Although a teacher's responses to the what, why and how are dependent of his professional knowledge, his responses are strongly impacted by his beliefs about mathematics or teaching and learning mathematics that are a part of the teachers' mathematical related affect (Hannula, 2012). For example, a teacher's beliefs are crucial for a teacher's decision to what extent he will enact his knowledge about mathematics and mathematics teaching and learning referring his instructional planning and, thus, for his classroom practice (e.g. Felbrich et al., 2012).

Accepting the impact of teachers' beliefs on both the instructional planning and the classroom practice, the further impact of teachers' beliefs, i.e. on the students' learning, seems obvious. However, although research in mathematics education yielded results referring to the relationships among teachers' beliefs on the one side, and the teachers' classroom practice and the students' learning on the other side (Artzt & Armour-Thomas, 1999; Staub & Stern, 2002; Dubberke et al., 2008), this relationships are not completely investigated (e.g. Hiebert & Grouws, 2007; Skott, 2009). In this report we concern two characteristics of teachers' beliefs that potentially could yield a consistency between teachers' espoused beliefs referring to their instructional planning and those beliefs that could be derived from classroom observations, i.e. the *centrality* of the expressed beliefs concerning the internal organisation of beliefs called belief system (Green, 1971; Putnam & Borko, 2000; Wilson & Cooney, 2002; Eichler, 2011; Schoenfeld, 2011), and the specificity of these beliefs referring to a mathematical subdomain (cf. Franke et al., 2007). The centrality of teachers' beliefs seems further to be crucial when professional development or, respectively, a change of teachers' beliefs is regarded (Borko & Putnam, 1996; Wilson & Cooney, 2002).

Our research concerns both, teachers' beliefs that are relevant for their classroom practice and the development of beliefs of teachers that we have followed from their final exams at university through a phase as teacher trainees up to their starting point as a qualified teacher. For this reason, a specific interest of our research and the focus in this report is to identify teachers' central beliefs restricted to the teaching and learning of arithmetic. According to this focus, we outline the theoretical framework and describe the method of our research. In addition to findings referring to individual arithmetic teachers' central beliefs, we discuss three types of arithmetic teachers (cf. Thompson, 1984). We conclude this report summarising our findings and suggesting further research.

2 Theoretical framework

Stein, Remillard and Smith (2007) provide a curriculum model including four phases of which the latter three phases are potentially influenced by teachers' beliefs (see fig. 1).

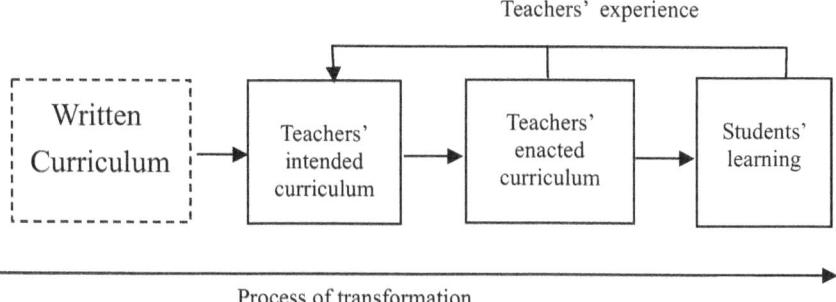

Figure 1 Four phases of the curriculum according to Stein at al. (2007)

The *written curriculum* involves instructional content, and teaching goals prescribed by national governments. The way the teachers interpret a written curriculum concerning content and goals referring to his instructional planning is called the *intended curriculum*. In this report, we mainly focus on teachers' intended curricula. However, indirectly we also regard the classroom practice involving interactions of a teacher with his or her students (*enacted curriculum*) and *students' learning*, since both have an impact on a teacher's intended curriculum through reflection on his or her experiences in classrooms (Wilson & Cooney, 2002).

We understand beliefs as an individual's personal conviction concerning a specific subject, which shapes an individual's way of both receiving information about a subject and acting in a specific situation (Pajares, 1992). Regarding this definition, we understand content and goals as specific forms of beliefs portraying a teacher's conviction about an appropriate way of teaching mathematics. Since an intended curriculum referring to arithmetic includes various specifications to appropriate content, goals or ways of teaching, we understand an intended curriculum as a specific form of a teacher's belief system (Green, 1971; Thompson, 1992). A belief system is characterised by a quasi-logical system of beliefs with different grades of centrality (Thompson, 1992). Although a teacher's belief system could potentially consists of clusters that need not to interact with each other, we hypothesise that a teacher's belief system is mostly consistent, if a specific mathematical domain, e.g. arithmetic, is regarded.

In this report we refer partly to overarching goals of the teachers that can be characterised by different features regarding the perception of mathematics in general (Dionne, 1984; Thompson, 1984) and to which Grigutsch et al. (1998) distinct four views:):

- A formalist view stresses that mathematics is characterised by a logical and formal approach. Accuracy and precision are most important.
- A process-oriented view is represented by statements about mathematics being experienced as a heuristic and creative activity that allows solving problems using different and individual ways.
- An instrumentalist view places emphasis on the "tool box"-aspect which means that mathematics is seen as a collection of calculation rules and procedures to be memorized and applied according to the given situation.

An application oriented view accentuates the utility of mathematics for the real world and the attempts to include real-world problems into mathematics classrooms. Further we refer to a global distinction of two different ways of teaching mathematics, i.e. a "cognitive constructivist orientation", and a "direct transmission view" (Staub & Stern, 2002, p. 344).

3 Method

The sample consists of 20 arithmetic teachers of primary and secondary school divided into two subsamples. The first subsample include 8 experienced teachers (four primary teachers, four secondary teachers) teaching arithmetic at least for five years. The second subsample consists even of 6 teachers (three primary teachers, three secondary teachers) that we have followed from their final exams at university through a phase as teacher trainees up to their starting point as a qualified teacher.

We collect data with a semi-structured interview including clusters of questions referring to arithmetic content, goals of teaching arithmetic, goals of teaching mathematics, the nature of mathematics, students' learning or materials used for the classroom practice, e.g. textbooks. In addition, the interviews incorporate prompts to evaluate given arithmetic tasks or fictitious statements of teachers or students that represent one of the views mentioned above, e.g. an application oriented view. Further, we used a questionnaire adapted form an existing scale referring to teachers' views (Grigutsch et al., 1998). We interviewed the experienced teachers once, since we assume these teachers' beliefs to be relatively stable (Calderhead, 1996). In contrast, we interviewed the teachers of the second subsample three times, i.e. at the end of their university studies, in the middle of their teacher training phase, and at the beginning of their time as a qualified teacher. The rationale for this longitudinal design is the assumption that prospective teachers' beliefs potentially change, when they get their first intense practical experience. These prospective teachers have little practical experience during their university studies including three internships that are mainly of observational nature. The teacher training between university and the beginning as a qualified teacher lasts 18 month and involves both self-dependent teaching and teaching guided by a mentor.

For analysing the data of the verbatim transcribed interviews, we used a qualitative coding method (Kuckartz, 2012) that is close to grounded theory (Glaser & Strauss, 1967). We used deductive codes derived from a theoretical perspective like "application oriented" goal and inductive codes for those goals we did not deduce from existing research concerning calculus education (Kuckartz, 2012). Further we weighted the codes with 1 or 2. If a teacher mentions a goal without a precision we weighted the code with 1. If a teacher explains a goal more deeply giving for instance a concrete example or task of his classroom practice, we weighted the code with 2. The codings were conducted by at least two persons and we proved the interrater reliability to show an appropriate value. Further, we analysed the sum of the weighted codes as triangulation to the qualitative interpretation of the interview transcripts. In a further triangulation we compared the results of the sum of weighted codes with the results referring to the questionnaire. We describe the results and the interpretation of the results of our method exemplarily in the next section referring to the structure of the belief system of one teacher.

4 Identifying central beliefs

In this section, we restrict the focus to one teacher, Mrs. A, and her beliefs system towards the teaching and learning of arithmetic. Referring to Mrs. A, we demonstrate three steps of analysis outlined above aiming to identify central beliefs in the belief system of a teacher. In the first step of analysis, we characterise a teacher's belief system on the basis of the interview transcripts.

> Mrs. A expressed coherently a process oriented view. That means, Mrs. A repeated her process oriented view on different parts of the interview. For example, to the question of her favourite teaching style and her preferred methods she answered:

Mrs. A: „Truly, it is important that they are able to find the solutions on their own, that they can work individually (…) that they can solve problems, that they can work on open tasks, that they can find their own strategies."

Later, nearly the same answer ensued when she was asked about pupils and their way of learning:

Mrs. A: „It is always important for me, that it comes from the pupils themselves, that it includes a problem, I like giving pupils problem statements."

Again, being asked to the question, which goals she would like to reach with her lesson, she answered:

Mrs. A: „And then there are strategies, i.e. to be flexible, to adapt oneself to something new. Therefore, you need the right attitude that you have the confidence to try something you don't know and to put effort into it."

The three quoted episodes referring to different topics, i.e. the teaching style, students' learning and teaching goals give evidence that beliefs representing the process oriented view are central in the belief system of Mrs. A.

According to the process-oriented beliefs Mrs. A expressed in various episodes of the interview she responded to prompts given during the interview. For example, Mrs. A was asked to arrange eight given teaching goals into a hierarchy. Figure 2 shows her arrangement of these goals for arithmetic lessons, where Mrs. A valued problem solving and process orientation as the most important goals.

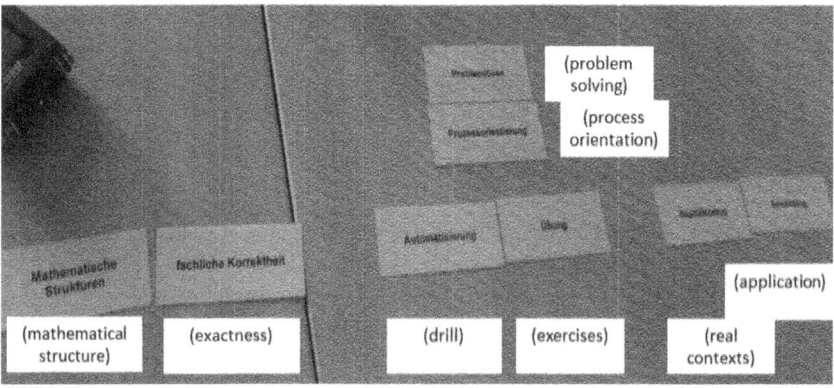

Figure 2   Mrs. A's arrangement of goals for arithmetic lessons

In figure 3 we show a further prompt consisting of students' statements representing the four views towards mathematics. The teachers were asked to arrange the statements from most desired (1) to least desired (4). Mrs. A preferred the second statement representing the process orientation.

---

8. statements of pupils:

[2] I like maths because there is a connection to real life problems.

[1] I like maths because hard nuts must be cracked and difficult problems can be solved.

[4] I like maths because many exercises can be solved by similar procedures/patterns.

[3] I like maths because the logic is clear and it follows strict mathematical rules.

Figure 3 Prompt: What would you like for pupils to answer?

Just as the espoused beliefs the responds to prompts give strong evidence that process orientation is central for Mrs. A.

In the second step of analysis, we coded every episode of the interview transcript. Referring to the deductive codes, partly given by views (application (A), formalism (F), process (P) and instrumentalism (I) and weighted the codes (see above). The sum of weighted codes is shown in figure 3 on the left side.

I the third step, the teachers were asked to complete a questionnaire according to the scale of Grigutsch et al (1998) and consisting a five-point-Likert-scale including 24 items representing the four mentioned view towards arithmetic (fig, 4, left side). To compare the weighted codes and the scores gained through the questionnaire, we standardised the sums of weighted codes and the questionnaire scores, which are both shown in figure 3 on the right side. Even for the individual teacher, we preliminary proved the fit of both distributions using correlation and ICC that show a good fit.

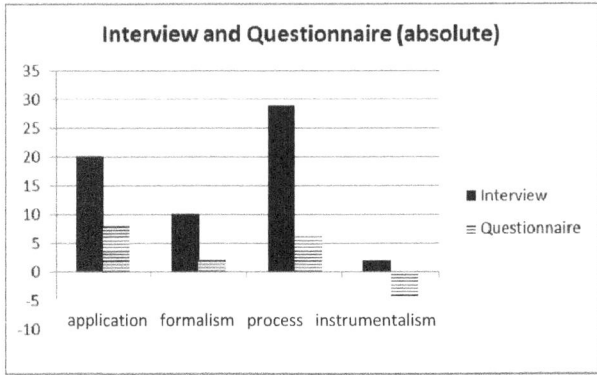

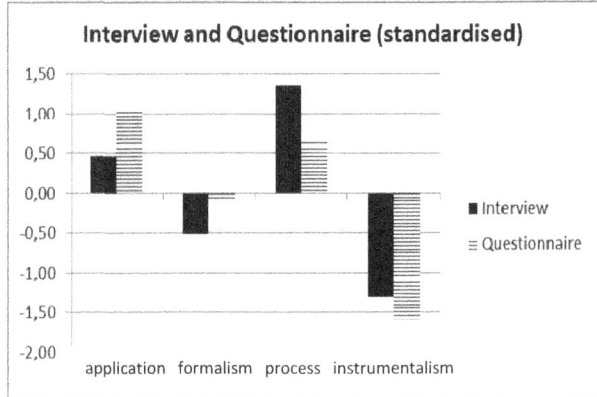

Figure 4   Weighted sum of codes and questionnaire scores

Concluding the analysis referring to the beliefs of Mrs. A concerning the teaching and learning of arithmetic, there exist several unambiguous examples for Mrs. A's process oriented view. The high degree of coherence in different parts of the interview, the sum of weighted codes and, finally the questionnaire underline the mentioned assumption that the process oriented view is central in the belief system of Mrs. A.

5   Explaining peripheral beliefs

Since the sum of weighted codes and the results of the questionnaire facilitate the identification of central and more peripheral beliefs, the interview transcripts provide a deep insight into the relationships of central and peripheral beliefs and also into primary and derivative (subordinated) beliefs (Thompson,

1992). For example, next to the process oriented view Mrs. A emphasised the importance of application (see fig. 3). Although application oriented goals are central for Mrs. A, however, her answers concerning the application oriented view give evidence that application oriented goals are subordinated to process oriented goals. Subordination means that that application is in some sense a central teaching goal but rather a means to an end for another even central and primary goal:

> „The relation to reality is important too, as I said before referring to money and time, but it doesn't have to be highlighted all the time. Today, for example, I just gave them a mathematical problem..."

This example shows that teachers can hold central beliefs that represent different views. In such a case we analyse relationships among the different views that were described exemplarily by regarding Mrs. A.

Concluding the analysis of the belief system of Mrs. A: On the one side, the sum of weigthed codes fit the results of the questionnaire and allows central and peripheral beliefs to be distinguished. On the other side, interpretation of the transcript allows to reconstruct the relationship of different central beliefs in terms of primary and subordinated beliefs and to explain beliefs as in detail based for example on specific tasks of a teacher's classroom practice.

6 Characterisation of teachers' belief systems

We restrict our focus to six teachers of our sample who were completely analysed yet. These six arithmetic teachers could be described by three views: Three teachers emphasise process-orientation and two emphasise application-orientation. The sixth teacher highlights partly the instrumentalism view and shows primarily a negative view towards process-orientation. Figure 5 summarises the findings for the teachers representing the three types of views in all three steps of analysis. The analysis of the interviews (column 1) shows the teachers' central beliefs, column 2 and 3 show additionally by the quotation of the interviews and questionnaires the matching of qualitative and quantitative results.

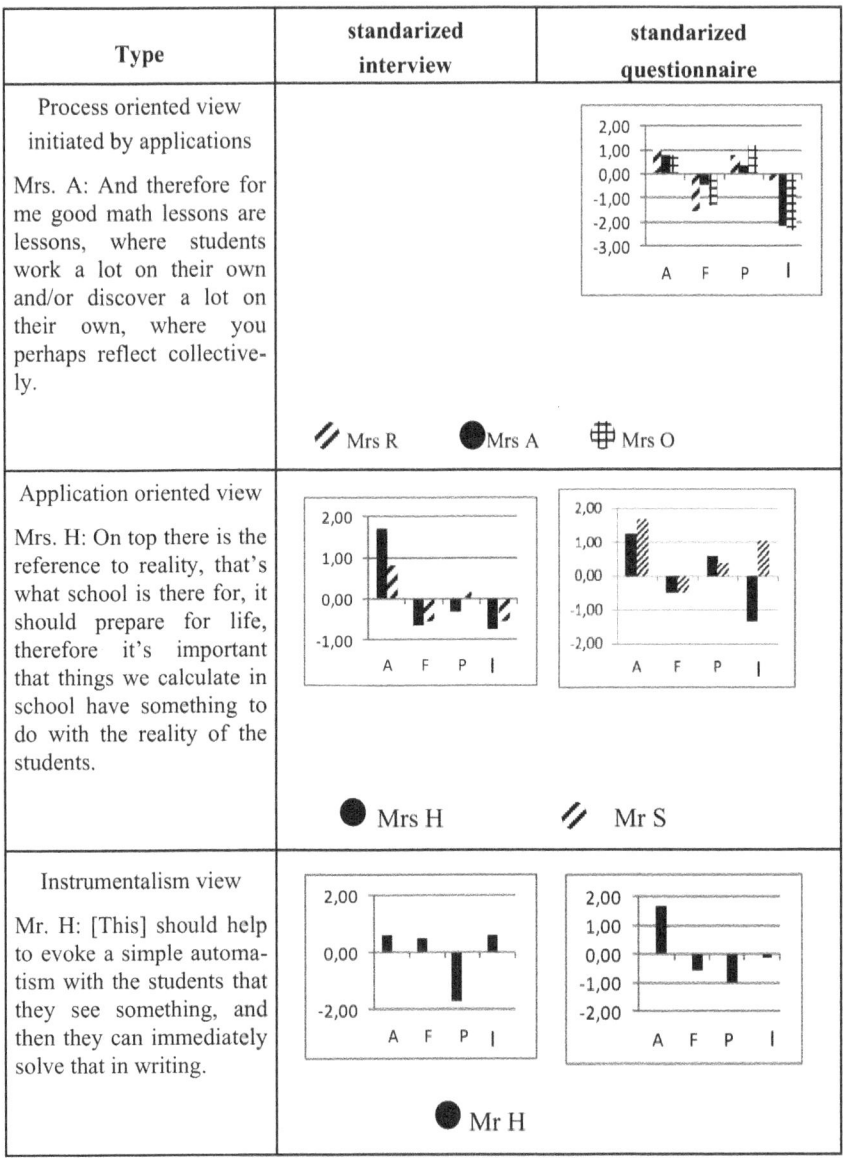

Figure 5   preliminary typing of the six teachers

It is striking that for all teachers the application oriented view is crucial and central for teaching arithmetic. However, the status of the application oriented view varies: For type 1 (e.g. represented by Mrs. A) the application oriented view is subordinated to the process oriented view, i.e. teachers of this type tend to use applications by means of achieving a process orientation. In contrast, for type 2 real-world problems are per se a crucial part of arithmetic teaching without emphasising a process oriented view. Finally, the teacher representing type 3 highlight applications particularly as a principle of student motivation. However, this teacher tends to subordinate the application oriented view to the instrumentalism view, i.e. applications are used to initiate arithmetic procedures the students have to learn. We illustrate the goal of student motivation referring to one episode of the interview with Mr. H:

> Mr. H: „It's commonly said you should pick things from the students' everyday life. This is crucial for initiating a subject and this is totally different compared to initiating a subject without these things. For example, if I have a family with 5 people and 3 pizzas, the students know that and can empathize with that and can perhaps understand the problem more easily. That's why problems should be from the students' everyday life."

7 Discussion

In this report, we presented a method aiming to identify teachers' central beliefs, which could further be distinguished to primary and subordinated beliefs, and peripheral beliefs constituting the teachers' belief systems towards teaching arithmetic. Results show that the qualitative interpretation of the interview as well as the weighted sum of codes and, finally, the analysis of the teachers' responds to a questionnaire consisting an existing scale (Grigutsch et al., 1998) yield consistent results referring to central and peripheral beliefs and, further, the qualitative analysis yield a distinction between primary and subordinated goals. We discussed these three steps of analysis in detail referring the process orientation of Mrs. A.

Taking into account the analysis of Mrs. A and further teachers, it was possible to analyse the status of peripheral beliefs. Thus, these peripheral beliefs are subordinated to central beliefs, i.e. the teachers express peripheral goals as a means to achieve central teaching goals. For example, applications are used to initiate problem solving (process oriented view; type 1) or to motivate students to learn arithmetic procedures (schema view; type 3).

On the basis of identifying central and peripheral beliefs as well as primary and subordinated beliefs, it was further possible to match the teachers to different types of teaching arithmetic. Partly, these types agree with the findings of Thompson (1984). In contrast to Thompson (ibid.), we found an omnipresence of favouring applications for the teaching of arithmetic.

We assume the central beliefs being relevant for both the teachers' observable classroom practice and the development of teachers' beliefs. Thus, the identification of the beliefs' characteristics serves as requirement for further steps in our research programme. In these steps, we will prove the relevance of central and peripheral beliefs and primary and subordinated beliefs by an observation of the classroom practice and the longitudinal analysis of teachers' beliefs.

8 References

Artzt, A. F. & Armour-Thomas, E. (1999). A cognitive model for examining teachers' instructional practice in mathematics: A guide for facilitating teacher reflection. *Educational Studies in Mathematics 40*, 211-235.

Borko, H., & Putnam, R. T. (1996). Learning to Teach. In D. C. Berliner & R. C. Calfee (Eds.), *Handbook of Educational Psychology* (pp. 673-708). New York: Macmillan.

Calderhead, J. (1996). Teachers: Beliefs and knowledge. In D. C. Berliner (Ed.), *Handbook of Education* (pp. 709-725). New York: MacMillan.

Dionne, J. J. (1984). The perception of mathematics among elementary school teachers. In J. M. Moser (Ed.), Proceedings of the 6th conference of the North American chapter of the international group for the psychology of mathematics education (pp. 223–228). Madison: University of Wisconsin-Madison.

Dubberke, T., Kunter, M., Julius-McElvany, N., Brunner, M. & Baumert, J. (2008). Lerntheoretische Überzeugungen von Mathematiklehrkräften: Einflüsse auf die Unterrichtsgestaltung und den Lernerfolg von Schülerinnen und Schülern. *Zeitschrift für Pädagogische Psychologie*, 22, 193–206.

Eichler, A. (2011). Statistics teachers and classroom practices. In C. Batanero, G. Burril, & C. Reading (Eds.), *Teaching Statistics in School Mathematics-Challenges for Teaching and Teacher Education* (pp. 175-186). New ICMI Study Series, vol. 15. Heidelberg, New York: Springer.

Felbrich, A., Kaiser, G. & Schmotz, C. The cultural dimension of beliefs: an investigation of future primary teachers' epistemological beliefs concerning the nature of mathematics in 15 countries. *ZDM – The International Journal on Mathematics Education* 44(3), 355-366.

Franke, M. L., Kazemi, E., & Battey, D. (2007). Understanding teaching and classroom practice in mathematics. F. K. Lester (Ed.), *Second handbook of research on mathematics teaching and learning* (pp. 225-256). Charlotte, NC: Information Age Publishing.

Green, T. (1971). *The Activities of Teaching*. New York, NY: McGraw-Hill.

Grigutsch, S., Raatz, U., & Törner, G. (1998). Einstellungen gegenüber Mathematik bei Mathematiklehrern. *Journal für Mathematikdidaktik 19* (1), 3-45.

Hiebert, J., & Grouws, D. A. (2007). The effect of classroom mathematics teaching on students' learning. In F. K. Lester (Ed.), *Second handbook of research on mathematics teaching and learning* (pp. 371-404). Charlotte, NC: Information Age Publishing.

Hannula, M. S. (2012). Exploring new dimensions of mathematics-related affect: embodied and social theories. *Research in Mathematics Education* 14 (2), S. 137–161.

Kuckartz, U. (2012). *Qualitative Inhaltsanalyse. Methoden, Praxis, Computerunterstützung.* Weinheim/Basel: Beltz Juventa.

Oliveira, H. & Hannula, M.S. (2008). Individual Prospective Mathematics Teachers: Studies on Their Professional Growth. In K. Krainer, & T. Wood (Eds.), *Participants in Mathematics Teacher Education*, Rotterdam, Sense Publishers.

Pajares, M. F. (1992). Teachers' beliefs and educational research: Cleaning up a messy construct. *Review of Educational Research*, 62(3), 307-332.

Putnam, R. T., & Borko, H. (2000). What do new views of knowledge and thinking have to say about research on teacher learning? *Educational Researcher, 29* (1), 4-15.

Schoenfeld, A. (2011). *How we think—a theory of goal-oriented decision making and its educational applications.* New York: Routledge.

Skott, J.(2009). Contextualising the notion of 'belief enactment'. *Journal for Mathematics Teacher Education* 12, 27-46.

Staub, F., & Stern, E. (2002). The nature of teacher's pedagogical content beliefs matters for students' achievement gains: quasi-experimental evidence from elementary mathematics. *Journal of Educational Psychology, 94* (2), 344-355.

Stein, M. K., Remillard, J., & Smith, M. S. (2007). How curriculum influences student learning. In F. K. Lester (Ed.), *Second handbook of research on mathematics teaching and learning* (pp. 319-369). Charlotte, NC: Information Age Publishing.

Strauss, A. & Corbin, J. (1996). *Grounded Theory.* Weinheim: PVU.

Thompson, A.G. (1984). The relationship of teachers' conceptions of mathematics and mathematics teaching to instructional practice. *Educational Studies in Mathematics* 15, 105-127.

Thompson, A. G. (1992). Teachers' beliefs and conceptions: A synthesis of the research. In D. A. Grouws (Ed.), *Handbook of research on mathematics teaching and learning* (pp. 127–146). New York: Macmillan.

Wilson, M. & Cooney, T.J. (2002). Mathematics teacher change and development. The role of beliefs. In Leder, G.C., Pehkonen, E.,Törner, G.(Eds.), *Beliefs: a hidden variable in mathematics education?* (pp. 127-148). Dordrecht: Kluwer.

We Think so, Me and My Mother - Considering External Participation Inside Teacher Education

Andreas Ebbelind

Linnaeus University, Sweden

Content

1 Introduction 110
2 Methodological and Theoretical Framing 110
   2.1   System Functional Linguistic 111
   2.2   Patterns of Participation 112
3 Methods Concerning the Study 113
4 The Case of Evie 113
5 Analysis 115
   5.1   Field and Ideational Meta-Function 115
   5.2   Tenor and the Interpersonal Meta-Function 117
   5.3   Mode and the Textual Meta-Function 117
6 Summary the Case of Evie 118
7 Discussion 119
8 References 120

Abstract

The aim of this paper is to show how external influences need be considered when discussing the formation of a primary school mathematics teacher. The external participation will be illustrated by the case of Evie, a student teacher. Two conceptual frameworks have been used, System Functional Linguistics and Patterns of Participation. The first has been used as a methodological tool and the second as an analytical tool. The results show that Evie's external prior and present participation might have an impact on her process of becoming a primary school mathematics teacher inside teacher education.

Keywords: Participation, Patterns of Participation, System Functional Linguistics, Teacher Education

1 Introduction

Many researchers have studied the relationship between student teachers image of teaching mathematics and their knowledge or experience gained inside teacher education (Phillip, 2007; Sowder, 2007). In this paper the experience that can be negotiated inside teacher education are referred to as internal participation. A central issue in this paper is that a student teacher brings multiple experiences into teacher education. Some of them derive from within mathematics education and others do not (Skott, 2013). Therefore this paper focuses on experience that stem from outside teacher education, but that are re-negotiated within it, so called external participation.

It is suggested that research on teacher development generally has focused on the individual paradigm, the learning-as-acquisition metaphor (Lerman, 2000; Sfard, 2009). However, in recent years the focus has been changing towards more social theories (Morgan, 2010; Phillip, 2007; Skott 2013). This paper follows the social and participatory paradigm, learning-as-participation metaphor (Lerman, 2000; Sfard, 2009), when interpreting data.

There are two objectives with this paper. First to present and use the methodological tool System Functional Linguistics (Halliday & Hasan, 1989), henceforth SFL, with the intention to unfold and disentangle the student teacher Evie's participation in prior and present practices. Secondly, the conceptual framework Patterns of Participation (Skott, 2013), henceforth PoP, will be used to interpret and illustrate the story of Evie, a student teacher. The aim is not to characterize external relevant practices, the aim is to explore and illustrate how a possible external influence are re-enacted and re-negotiated, inside teacher education, in the process of becoming an upper primary school mathematics teacher.

2 Methodological and Theoretical Framing

Two conceptual frameworks are used as methodological and analytic tools. First to address and unfold situated communication SFL has been used (Halliday & Hasan, 1989). It illustrates linguistic choises as results of prior and present participation (Meaney, 2005). However, SFL does not emphasise any mathematical content. In the case presented in this paper it mainly focuses on how the student teacher, during interviews, addresses the content or situation. Secondly, the conceptual framework PoP will be used when interpreting the data (Skott, 2013). The outlined approach therefor uses SFL to unfold situated communication to reveal traces of context. These traces of context are then interpreted through PoP.

2.1 System Functional Linguistic

In SFL a text is something that is constructed while participating with others. It consists of everyday language and specific terms that are focus the producer's awareness of the context of situation. The context of situation is the environment of the text itself and is described through field, tenor and mode (Halliday & Hasan, 1989). These concepts, which are described below, serve to interpret the social context of a text (Halliday & Hasan, 1989. A text is handled in three different processes, so-called meta-functions, simultaneously. Morgan (2006) points out that this unfolding into meta-functions serves as a crucial window when following processes, in this case the process of becoming a primary school mathematics teacher. I will present these meta-functions together with the notions of field, tenor and mode.

The field concerns what is going on? The first meta-functions, the *ideational meta-function*, addresses peoples experience in some kind of process and is realised through the field (Morgan, 2006; Schleppegrell, 2007). It concerns verbs and how processes are expressed through the so-called transivity-system, material, relational, verbal and mental processes. Material processes involve physical actions. Relational processes emphasise relations between objects. Verbal processes express something that has been said and mental processes addressing phenomenon.

The tenor highlights the participants and what choices they have according to power relations, status and roles. The second function is called the *interpersonal meta-function* and is realised through the tenor (Morgan, 2006; Schleppegrell, 2007). This meta-function concerns the tense, prior and present participation, and to what extent the proposition/clause is valid. This is more active than the ideational meta-function; it is interpersonal meaning that it is both interactive and personal and highlights the choices the participants have in the situation.

Finally we consider the context and the language surrounding us. This is done within situations and calls the *textual meta-function*. This function is realised through the mode (Meaney, 2005; Morgan, 2006). It concerns the process of construing the coherence of a text. The approach described is illustrated in Table 1.

Table 1: The methodological tool

Situation/Discourse	Realised by	Meta-function
Field ✓ What is going on? ✓ What are the participants engaged in?	**Ideational Function** Transivity Naming	 Material processes: There is an actor (doer) that does something. Mental processes: The senser is addressing a phenomenon. Relational processes: Emphasise relations between objects. Verbal processes: Express something that has been said. Concerns the naming of objects that is evident in the linguistic choices made by participants.
Tenor ✓ Who are taking part? ✓ What entities are visible? ✓ What choice according to power/status/role?	**Interpersonal Function** Tense Polarity Modality Voice	 Present, past or future time. Positive or negative validity. To what extent the proposition is valid. Imperative and personal pronouns.
Mode ✓ What role are assigned to discourse/language?	**Textual Function** Cohesiveness Lexical chain	 Relation to the Context of Culture Being cohesive to the subject.

2.2 Patterns of Participation

PoP draw according to Skott (2013) on two main theoretical sources, symbolic interactionism and social practice theory. In social practice theory student teachers' identity formation and learning are a result of shifted participation in educational situations. It acknowledges that all activities are situated. The aim in PoP is to understand how a persons interpretations of and contributions to immediate social interactions in interviews relate to prior engagement in a range of other social practices.

Teacher education requires participants, that is, individuals who participate and negotiate meaning when positioning oneself in practice. The individual student teacher is part of a situation and since participation is situated in a specific location individuals participate in specific ways. A student teacher brings multiple ways of participating into different contexts of situations that constitutes teacher education. Some of them derive from mathematics education and others do not (Skott, 2013).

In PoP student teachers' individual mathematical skills or beliefs are not the object of inquiry, but the process said to precede so-called mental construct. Expressed in another way, it describes differently what is usually described as mathematical skills, knowledge or beliefs. PoP are phrasing, in participatory terms, the shifted movement when relating to different aspects of teaching and learning mathematics. In other words, it intend to use processual interpretations when describing the process of becoming a mathematical teacher, in this case a primary school matematics teacher.

3 Methods Concerning the Study

The present study adopts a theory driven, multi-site ethnographic approach. It involved six student teachers and followed them before, during and after different situations such as lectures, seminars, internships, study groups and examination work. The study was theory driven (Walford, 2009), because theories that emphasise situatedness (Lave & Wenger, 1991; Skott, 2013) guided the choices made during the on-going project. The study was multi-sited because the mode of construction was not a single site; instead the mode of construction was a process that took place in multiple sites (Pierides, 2010).

The six student teachers were chosen to ensure that the cases were critical (Flyvbjerg 2006). They are selected for their commitment and their mathematical knowledge. "In turn, this may allow novel interpretations and analytic generalisations about relationships between [student] teachers' participation [...] on the one hand and in educational practices and discourses beyond it on the other." (Skott, 2013, p. 8)

This paper concern one student teacher and is presented as narrative, the story of Evie, and intends to capture the student teachers evaluations and storie about herselves as teacher-to-be. To be more specific, it inteds to capture the shifts in Evie participation expressed during three semi-structured interviews (Kvale, Brinkmann & Torhell, 2009).

A text is any instance of language being used as part of a context of a situation. Therefore, every text reflects that it is about something, is addressed to someone and uses a particular mode, spoken or written language for example, to express its meanings. In this paper short episodes that reflect the external context of situation related to Evie's mother are chosen. These episodes are then analysed by SFL and interpreted through PoP.

4 The Case of Evie

Evie was included in this study because of her interest in mathematics and because she intended to write at least one master thesis, 15 credits, at advanced level in mathematics education. This means that she would have at least 45 credits in mathematics education after her graduation (total 240 credits during four years). However, the aim is not to focus on Evie as such but to illustrate how external influences needs to be considered when discussing the formation of a primary school mathematics teacher. The texts below are from three interviews with Evie and one interview with Evie's mother.

Evie, who was in her 20s, started teacher education directly after high school. According to Evie she came from a family history where mathematics was the main school subject. Her family was proud that she wanted to continue studying after high school, especially the choice of going into teaching. The first interview was conducted at the very beginning of her teacher education programme and focused on her prior classroom experiences and her current understandings about teaching and learning mathematics. Evie emphasised that being a part of this study was her mother's suggestion. She believed it was good for Evie to talk about educational things with a researcher.

> Evie: When you talk, you learn. [...] We talk a lot about school at home, because my mother is a teacher. We think mathematics education is great. I come from a mathematics family and everybody is interested in mathematics.
>
> Researcher: You mentioned that your mother was a teacher, has she any influence on you?
>
> Evie: No, I do not think so.
>
> Researcher: But you have some thoughts about teaching!
>
> Evie: Yes, maybe we talk sometimes, but on the same time you want to change you own experience. [...] I want to say that I know what it means to be a teacher because I have a mother and my father's sister that are teachers. I know what I engage in.

The second interview was made at the end of the first five-week internship, after approximately seven months of university-based courses. It concerned Evie's experiences with teacher education programme as they relate to what mathematics is, to mathematics learning and teaching and to what it means to know mathematics.

> Evie: [Mathematics is] one part of the society and a basic knowledge so one can manage, above all, when shopping and when something is needed in every day life. Not just something that one shall do.
>
> Researcher: Ones again you come back to usefulness. Have you always thought of it in this way?
>
> Evie: It derives from my own schooling and we have talked a lot about it at home.
>
> Researcher: Your mother, does she think she is successful in communication usefulness?
>
> Evie: Yes I think so; she tells that many students ask questions about if this is useful. She has made that clear to them.

The third interview with Evie was conducted one week after entering a 30 credits, 20 weeks, course in mathematics education, approximately 18 months after Evie began her studies at university.

> Researcher: How much do you [Evie and her mother] discusses your schoolwork?
>
> Evie: Well, she wants the material. But that is more for her own teaching, all exercises and booklets. I will give her advice what to do. We do not discuss my schoolwork, but I send her my tasks [assignments] and ask if they are okay.
>
> Researcher: Do you disagree with her sometimes?
>
> Evie: No I agree to the critic and change.
>
> Researcher: So you are sending your assignments to her for response?
>
> Evie: Yes, especially during the history course. Sometimes I feel that my thoughts aren't ready, I need input on what I might have forgotten.

One interview was conducted with Evie's mother, Angela, an upper primary school teacher who specialised in mathematics education. Angela was interviewed at her work place after Evie's first year of studies. Important notions for her in relation to teaching mathematics were, learning the basics, practical work, *"and of course you need rote learning"*, group work, and reality-based learning. She was very critical to her own teacher education programme from 2005, which, according to her, lacked any element of how to teach. Therefore she considered it *"a great opportunity"* for Evie to get the quality education that she did not get herself. At the end of this interview Angela concluded that Evie's education is good for her to.

5 Analysis

In this section the situated communications presented will be analysed through the context of situation that constitutes the meta-function.

5.1 Field and Ideational Meta-Function

The first meta-function concerns people addressed experiences and are constituted by the field. Firstly there is a situation, a conversation with two persons. In this situation the researcher poses questions, imperatives, and guides the conversation. The student teacher makes, in the situation, linguistic choices to present an evaluation when engaging and re-engaging in prior, present and future practises.

In the first interview Evie indicates that school is a common subject of discussion and identifies her family in relation to mathematics. In the second interview she engages in a discussion about why mathematics is important. The researcher asks her. *"Ones again you come back to usefulness. Have you always thought of it in this way?"* Evie answers this question by re-engaging in a discussion about her mother. In the last interview she positions herself in relation to her mother. She engages in a discussion about the role that her mother takes in relation to her education.

The ideational meta-function addresses peoples experience in some kind of process. In the transcript from the first interview Evie addresses material processes. These are *"change"*, *"be"* and *"have"*. She is visible in all these processes as a doer. The first refers to Evie's own experience and the last two addresses the choice of entering teacher education. *"I want to say that I know what it means to be a teacher because [...]."* The verbal process *"talk"* are used two times and addresses her family and *"say"* is used to emphasize her claim of being aware of her choice. The mental processes in the text refer to *"think"* and *"know"*. The verb *"think"* refers to the family as one unit, *"we"*. There is one relational process, *"come"*. It describes the relation between herself, her family and their knowledge in mathematics.

In the second interview the processes refers to mathematics, earlier school experience, the relation between Evie and Angela and Angela's day-to-day work. Evie re-engages in their prior conversations about Angela's practice and in their mutual interest in teacher education. The material processes *"shopping and do"* refer to the role mathematics play in society and *"has"* to the use of mathematics. *"[T]alk"* and *"tells"* refer to the relation between mother and daughter. There are three relational processes, *"manage"* and *"needed"* is related to mathematics in society. *"Derives"* is related to the usefulness and prior school experience. There is one mental process in this episode, *"think"*, and address Angela's day-to-day work.

In the last interview the material processes are *"give"* and *"send"*. Evie addresses two different practices, first the *"give"* refers to her mother's need for teaching material and secondly *"send"* to their silent communication regarding her assignments. The mental processes in the text refer to *"agree"* and *"feel"*. When she addresses these processes it might concern her mother's authority as a teacher or even as a mother. *"I agree to the critic [written response from Angela] and change"*. There are two verbal processes in this episode, *"discuss"* and *"ask"*. There are no relational processes, in this text the participation is entirely on the mothers terms.

5.2 Tenor and the Interpersonal Meta-Function

The second meta-functions, interpersonal, concerned whom the addressed experience refers to and are constituted by the tenor. There are two persons present in the situation, the researcher and the student. Evie has in the first and second interview addressed prior and present experiences and brought her family and mother into the conversation. In the last interview the researcher guide the linguistic choices to reveal who helps her with her academic work. In this section the power relation is that the researcher's questions act like imperatives. Once again Evie includes her mother into the situation. Angela is taking part through Evie's re-engagement in prior participation. It is clear that the power relation between mother and daughter are explicit in this interview.

The second function is more active than the ideational meta-function; it is interpersonal, i.e. both interactive and personal. The personal pronoun *"I"* is important in relation to prior experience in all interviews. In the first interview she re-engages with her family and her mother. On one occasion Evie refers to future practices, she *"want[s] to change"* otherwise she uses past and present tense. This change is in relation to her prior experience and to her mother's practice. She positions herself as someone that not wants to completely repeat the teaching she experienced in school. In the second interview *"we"* refers to Angela and Evie. There is another central actor in this interview, Angela's practice interpreted by Evie. When Evie uses *"They"* it refers to Angela's pupils and how they experience her practice. In the last interview Evie, her mother, and the teacher education programme are visible.

5.3 Mode and the Textual Meta-Function

The third meta-functions, textual ones, concern cohesive relations when adapting to the surroundings and are constituted by mode. In all interviews language is used to evaluate understanding of topics that are related to mathematics education. Spoken language acts like a tool when Evie engages in prior, present and future practice.

The lexical chain is intact in all three interviews. The story is centred through the "I" and getting its strength through other participants, family, Angela and Angela's practise. The function of these is to visualise that this is not only my, Evie's, thoughts. In the second interview "we" refers also to Angela's proven practice. The use of the conjunction "because", in the first interview, is to convince the listener. Finally in the third transcript the conjunction "but" is indicating a contradiction. It might be that Evie is expecting a coherent relationship with her mother but there are no indications in the textual meta-function that this is the case.

6 Summary the Case of Evie

The researcher was a participant who used imperatives to guide the interview. Evie was the participant who was interviewed and she brought her mother in as an active participant, especially her mother's practice. From the linguistic choices it seems that there is a change of character in Angela and Evie's relation. Angela is changed, merged or constructed to an authority that Evie is trusting to have the right interpretation regarding teacher education. A relevant question in this case is weather this relationship complements, constrains or constructs her understanding of what mathematics is, how to learn and teach mathematics, and what it means to know mathematics.

During the three interviews Evie was evaluating her prior and present participation. Through her linguistic choices she negotiated meaning and re-engaged herself in relation to other relevant practices. She chose to present pieces of her prior, present or future practices. The word "we" is central when interpreting what is going on. My interpretation is that her family, especially her mother, complement the teacher education. Evie re-enacts and re-negotiates her experience gained inside teacher education in relation to her family and especially her mother. Her experience from this external and internal participation merges and constructs her current understanding of teaching and learning mathematics.

During time of the study Evie's linguistic choices became more and more content specific in relation to what mathematics is, how to learn and teach mathematics and what it means to know mathematics. In the first interview "we" refers to her family, all members, and her mother as parts of her current pattern. When she in the second and third interview mentioned "we" and "family" it referred to herself and her mother. There is another small active part visual in the second interview and central in the last one. Evie is re-engaging herself in Angela's teaching and therefore her mother's teaching becomes visual. Her evolved understanding helps her become more content specific when addressing Angela's outspoken teaching. Angela's teaching becomes a significant linguistic choice, an aspect of Evie's current PoP.

Evie's relationship to her mother changed when referring to teacher education. From a person that you talk with, to a person to whom you owe gratitude, and finally to a person who has the right interpretation and knowledge. It also seems as if they have gone from a more mutual communication to a more one-way directed communication. The external participation becomes important when negotiating the internal participation.

In the interview Evie's obviously used spoken language to enact and re-enact prior and present experiences. She told the story of Evie, but this story got its inspiration from other external participation. By re-engaging in external practises the story tells us that there are not only Evie's thoughts. There are thoughts of proven practice behind her evaluations. Central aspects of Evie's Patterns of Participation are re-enacted from the outside when she positioned herself as a student teacher inside teacher education.

7 Discussion

This paper promotes a participatory interpretation of student teachers' development during teacher education. The approach is to consider not only teacher education itself but also other external types of participation. It is interested in teacher students shifted participation during their evaluations of what mathematics is, how to learn and teach mathematics, and what it means to know mathematics. A weakness in this paper is that it only contains one case and one external influence, whereas the overall study is a multiple-case study that can show variations and possible similar prior engagement. But the aim was not to characterize external relevant practices.

The aim was to investigate and illustrate, through PoP, how external influences are possibly re-enacted and re-negotiated when a student teacher is becoming a primary school mathematics teacher and to raise some points in the interpretation that can be considered.

The methodological tool System Functional Linguistics (Halliday & Hasan, 1989) made a fine-grained unfolding possible that made the process of becoming an upper primary school teacher visible (Morgan, 2006). It disentangled the student teacher Evie's participation in prior and present practices. This made the interpretation and illustration of Evie's story possible using Patterns of Participation. SFL is used for unfold the meaning in texts and this can lead to relevant interpretations of how external practises operate in these texts.

The story of Evie is my interpretation of the linguistic features that are visible in the transcripts. It is a story that draws on the verbs, the process, and the entities that it refers to. The material, mental, verbal and relational processes are described in relation to the participants that are visible. Of course I might be constrained by my own background, but I consider my interpretation of the linguistic features valid and hope that it has raised some points or contributed to thoughts about the complexity of becoming a teacher.

This paper highlights the external participation that might influence the student teachers becoming. It shows that we need to consider not only the teacher education itself but also understanding in how other relevant practices contribute to student teachers understanding.

8 References

Halliday, M. A. K., & Hasan, R. (1989). Language, context, and text : aspects of language in a social-semiotic perspective (2. ed.). Oxford: Oxford Univ. Press.

Flyvbjerg, B. (2006). Five misunderstandings about case study research. Qualitative inquiry, 12(2), 219–245.

Lave, J., & Wenger, E. (1991). Situated learning : legitimate peripheral participation. Cambridge: Cambridge Univ. Press.

Lerman, S. (2000). The social turn in mathematics education research. In J. Boaler (Ed.), Multiple Perspectives on mathematics teaching and learning (pp. 19-44). Westport, CT: Ablex.

Meaney, T. (2005). Mathematics as text. In A. Chronaki & I. M. Christiansen (Eds.), Challenging Perspectives on Mathematics Classroom Communication (pp. 109-141). Greenwich, Conn.: IAP-Information Age Pub.

Morgan, C. (2006). What does social semiotics have to offer mathematics education Research? Educational Studies in Mathematics, 61(1), 219-245.

Phillip, A. R. (2007). Mathematics teachers beliefs and affect In F. K. Lester (Ed.), Second handbook of Research on Mathematics Teaching and Learning (pp. 257-318). Reston: National Council of Teachers of Mathematics.

Pierides, D. (2010). Multi-sited ethnography and the field of educational research. Critical Studies in Education, 51(2), 179-195.

Schleppegrell, M. J. (2007). The linguistic challenges of mathematics teaching and learning: A Research Review. Reading & Writing Quarterly, 23(2), 139-159.

Sfard, A. (2009). Moving between discourses: From learning-as-acquisition to learning-as-participation. AIP Conference Proceedings, 1179(1), 55-58.

Skott, J. (2013). Understanding the role of the teacher for emerging classroom practices: searching for patterns of participation. ZDM - The International Journal on Mathematics Education, 45(4).

Sowder, J. T. (2007). The matematical education and development of teachers. In F. K. Lester (Ed.), Second handbook of Research on Mathematics Teaching and Learning (pp. 257-318). Reston: National Council of Teachers of Mathematics.

Walford, G. (2009). For ethnography. Ethnography and Education, 4(3), 271-282.

Primary School Teachers' Image of a Mathematics Teacher

Hanna Palmér
Linnæus University, Sweden

Content

1 Introduction 122
2 Theoretical Framing 123
3 The Study 124
4 Results 125
    4.1   The Time of Graduation 125
    4.2   The Two Years after Graduation 126
    4.3   Group Interview Two Years after Graduation 127
5 Analysis 127
6 Conclusion and Discussion 129
7 References 130

Abstract

The results presented in this paper derive from a longitudinal case study of seven novice primary school mathematics teachers' professional identity development. In the study it was found that this professional identity development did not include becoming a mathematics teacher. A primary school teacher in Sweden, like in many other countries, teaches many subjects but, at the same time they are the first teachers to teach mathematics to the school children. In the paper it will be shown how the novice primary school teachers' image of a mathematics teacher prevented them from developing a sense of themselves as mathematics teachers.

1 Introduction

Research on teachers' professional identity formation has expanded in recent years with the mutual goal to better understand and support the needs of teachers, including student teachers (Beijaard, Meijer & Verloop, 2004; Bjuland, Luiza Cestari & Borgersen 2012; Ponte & Chapman, 2008). The results presented in this paper derive from a longitudinal case study of seven novice primary school mathematics teachers' professional identity development (Palmér, 2013). A teacher's professional identity is neither totally collective nor totally individual. A teacher is expected to have some characteristic professional knowledge, goals and attitudes but, at the same time, teachers are autonomous and differ with regard to knowledge, goals and attitudes (Beijaard, Meijer & Verloop, 2004). Further, the teaching profession is practiced in different contexts, which creates a plural teacher identity, including, for example, a mathematics thinker, a teacher in the classroom, a mentor for students, a colleague and so on. The unified professional identity is a teacher but it is practiced in different contexts in different communities, for example in the classroom and at meetings, and is therefore affected differently (Schifter, 1996; Sachs, 2001). Based on above, a teacher's professional identity is simultaneously individual, collective, plural and practiced in different contexts which is a challenge for the expanded research in the area.

According to McNally, Blake, Corbin and Gray (2008) the transfer from teacher education to teaching is to be seen as a shift in identity, where becoming accepted as a teacher by colleagues but also by oneself is central.

> Beginners in teaching face the fundamental question of whether they can see themselves as teachers, not only the reflections from colleagues and children in their schools, but also in the mirror that they hold up to themselves (McNally, Blake, Corbin & Gray, 2008, p.295).

In the here presented study of seven novice primary school mathematics teachers' professional identity development it was found that their professional identity development did not include becoming a mathematics teacher. Even if they taught mathematics they did not see a mathematics teacher "in the mirror that they [held] up to themselves" (McNally et al., 2008, p.295). When analysing the novice primary school teachers' professional identity development the first two years after their graduation, a question emerged regarding what it was that made them *not* think of themselves as mathematics teachers. That question is what will be focused on in this paper as *the image of a mathematics teacher.*

2 Theoretical Framing

According to Lerman (2000), research into mathematics education has "been turn[ed] to social theories" (p.20). He bases this on mathematics education research since the late 20th century, sees meaning, thinking and reasoning as products of social activities where learning, thinking and reasoning are seen as situated in social situations. The term situated refers to a set of theoretical perspectives and lines of research which conceptualise learning as changes in participation in socially organised activities and individuals' use of knowledge as an aspect of their participation in social practices (Borko, 2004). Peressini, Borko, Romagnano, Knuth and Willis (2004) argue for using such a situative perspective in studies of mathematics teachers' teaching.

According to Gee (2000-2001), identity is to be recognised (by oneself and/or others) as *a kind of person* in a given context, which would imply that professional identity as a mathematics teacher is being recognised (by oneself and/or others) as a mathematics teacher in a given context. As such, identity has both individual and social elements. To be recognised (by oneself and/or others) as *a kind of person* in a given context is neverending implying identity as a process. Similarly Morgan (2009) writes that establishing a (positive) professional identity as a mathematics teacher involves positioning oneself "within discourses of education in general and mathematics teaching in particular (p. 109)" in ways that allow one to be seen by others and oneself as a (good) teacher of mathematics.

In the study presented in this paper, two situative theoretical perspectives, communities of practice (Wenger, 1998) and patterns of participation (Skott, 2010; Skott, Moeskær Larsen, & Østergaard, 2011), are coordinated in a conceptual framework aiming to capture both the individual and the social part of identity development involved in the over-described recognition as *a kind of person*. According to Skott et al (2011) a teacher participates in "multiple simultaneous practices" (p.32) in the classroom and there are patterns in the ways in which the teacher participates in these practices. The aim in patterns of participation research is to understand how a teacher's interpretations of and contributions to immediate social interactions in the classroom relate to prior engagement in a range of other social practices. These other social practices are in the study treated as communities of practices by Wenger (1998). An individual's patterns of participation in different communities of practice influence how they are recognised, by oneself and others. To become recognised, by oneself and others, as a mathematics teacher, an individual's patterns of participation have to be in line with the patterns of participation of a mathematics teacher. But, what are the patterns of participation of a mathematics teacher?

3 The Study

The study of primary school teachers' professional identity development is a case study with an ethnographic approach, where seven Swedish novice primary school teachers have been followed from their graduation and two years onwards. In Sweden, primary school teachers most often work as class teachers teaching several subjects whereof mathematics may be one. This is similar to other countries around the world were most primary school teachers are educated as generalists (Tatto, Lerman & Novotná, 2009).

The teacher education the respondents were to graduate from at the beginning of the study is an integrated teacher education where professional and subject studies take place concurrently. Compared to the previous primary school teacher educations in Sweden, content knowledge was emphasised with decreased practice periods (Lindström Nilsson, 2012). The respondents in the here presented study were selected because they in teacher education chose mathematics as one of their main subjects. Some of them also wrote their final teacher education Bachelor theses on mathematics education. As a minimum the respondents in the study had taken 22,5 credits, at most 52,5 credits, of courses within the field of mathematics education. The meaning with this selection was to maximise the possibility for the respondents to teach mathematics after their graduation.

The ethnographic approach was used to make visible both the individual and the social part of professional identity development. The empirical material was collected through self-recordings made by the respondents, observations and interviews. To accomplish a balance between an inside and outside perspective (Aspers, 2007); the observations were both participating and non-participating. For the same purpose the interviews were both spontaneous conversations during observations and formal interviews (individual and in groups) based on thematic interview guides. These varying empirical materials have different characteristics but are in the analysis treated as complete-empiricism (Aspers, 2007) implying that all the empirical material constitutes a whole.

The results presented in this paper have been developed gradually based on interplay between fieldwork and analysis of observations, interviews and self-recordings. The starting point of the analysis in ethnographic research is the meaning the respondents themselves infer on the situations studied (Aspers, 2007; Hammersley & Atkinson, 2007). The analysis in this study has been done using grounded theory methods which implies building and connecting categories grounded in the empirical material by using codes (Charmaz, 2006). Coding the empirical material does not imply using pre-constructed codes, but labelling the empirical material, line-by-line, with as many codes as possible (Kelle, 2007). Based on the question what it is that make the respondents recognise themselves as a kind of mathematics teacher (or not) segments in the empirical material were inductive labelled with the codes attribute, image, criteria and epithet. After that, these codes were deductive connected through axial coding. Finally, through the writing of memos the category *the image of a mathematics teacher* emerged. This category is the similarities found in the unique experiences and expressions of the respondents regarding what it implies to be a mathematics teacher.

4 Results

This section will be presented in three sub-sections. In the first section the time of graduation will be focused on. The second section contains a summary of the respondents' two years after graduation. Finally, in the third section, parts from a group interview two years after graduation is presented. The joint theme in the sub-sections is the similarities of the respondents regarding what it implies to be a mathematics teacher. The empirical examples below are not to be seen as the wholeness of what the category *the image of a mathematics teacher* is based on but as examples of empirical instances labelled within that category.

4.1 The Time of Graduation

The respondents were interviewed the first time just before their graduation from teacher education. By other things they were asked why they wanted to become primary school teachers and why they had chosen mathematics as one of their main subjects. They were also asked if they had any mathematics teachers as role models.

The respondents' motives to become teachers were connected to interest in working with children and in school. The choice of mathematics as one of their main subjects was explained as either a tactical choice or as a choice of interest.

> I have always liked mathematics a really lot. At least until upper secondary school. And I loved it already when I started school. [...] mathematics is anyhow a subject I have always liked myself and I feel that it is fun, or like that. Then it is much easier. For example, quite the opposite I have had a really hard time with English and then you feel it is difficult to motivate the students to enjoy English when I have always experienced it as really, really difficult myself. (Camilla)

Mathematics as a tactical choice is explained as mathematics together with Swedish being the most important subjects in primary school.

> I love to work with children! I chose Swedish and mathematics for younger children since I think they are the foundation of the Swedish education system and they are the most important subjects. (Jenny)

> I guess I would have preferred to study physical education but since there are few physical education teachers in primary school, I thought it was cleverer to have science and mathematics when finished and applying for job. (Nina)

Only one of the respondents remembers having had a good mathematics teacher in school and that was in municipality adult education.

> I can't think of anyone because I don't think I have met anyone who, as I see it, is a really good mathematics teacher. There are mathematics teachers who have good knowledge of mathematics, but that doesn't mean that they can teach mathematics. (Barbro)

> I haven't had what I think is a good mathematics teacher. [...] It is not enough to explain the same thing ten times and think that the student will understand. I have experienced that many times, that they explain in the same way over and over again. (Gunilla)

In time for graduation all of the respondents express a clear opinion regarding how mathematics ought to be taught in primary school and they want to reform mathematics teaching. They say that they have met a new way to teach mathematics in their teacher education. This new way to teach mathematics differs from the mathematics teaching they themselves have experienced in school.

> I believe that there are many different ways today. When we went to school, you were only allowed to work in one way. Today there are different ways. (Barbro)

> The first time I saw a cubic meter and realised that I could fit inside it I was totally surprised. And I experienced that the first time here at the university. Why haven't you experienced that in your own schooling when you were little? (Gunilla)

4.2 The Two Years after Graduation

After graduation the respondents start to work at different schools and preschools as class teachers, long-term and short-term substitute teachers and as teacher assistants[1]. Some of them teach mathematics a lot, others more sporadic. However, in one way or another, all of them teach mathematics during the two first years after graduation. Similar for all of them is that they, even when they teach mathematics a lot, do not emphasis mathematics in their work. Not much is seen of the new way to teach mathematics that the respondents emphasized before graduation. Quite the opposite they teach mathematics in a way that they disaffiliated themselves from before graduation.

---

[1] While collecting the empirical material for this study (2009-2010), it was difficult for primary school teachers in Sweden to get jobs, especially in certain municipalities.

4.3 Group Interview Two Years after Graduation

Two years after their graduation from teacher education the respondents are gathered for a group interview. When, in that group interview, being asked if they feel like a mathematics teacher, they all say no. For some of them mathematics teaching is not included in their teacher assignment right now but nor those who work as class teachers, teaching mathematics every day, express having a sense of themselves as a mathematics teacher.

>Gunilla   It's easy for me to answer that question. I don't feel like a teacher of mathematics[2].
>
>Researcher   Have you felt like it at any time since graduation?
>
>Gunilla   No
>
>Researcher   Not even when you were teaching it?
>
>Gunilla   No. [...] But no, I don't feel like a mathematics teacher but I can absolutely see myself having a job as a class teacher within which teaching mathematics is a part.
>
>Researcher   Nina?
>
>Nina [...] I have quite a lot mathematics right now but my biggest dilemma is that I came in like that, and have to practice mathematics teaching that's already been started. And I may not be one hundred per cent. I can feel like the next time I have mathematics, if I'm on my own and am to use a text book. I'll choose the text book I want to use, and how to use it.

Further, Helena says that she would like to work more as a "subject teacher" and not have to bother about subjects she has not got in her teaching degree. She also says that she have to learn more about the history of mathematics to develop a sense of herself as *a kind of mathematics teacher.*

5   Analysis

When analysing the empirical material with focus on why the respondents did *not* think of themselves as mathematics teachers, segments were labelled with the codes attribute, image, criteria and epithet. These codes put together is the similarities found in the unique experiences and expressions of the respondents regarding what it implies to be a mathematics teacher, their *image of a mathematics teacher.*

When, in the group interview two years after graduation, being asked if they felt like a mathematics teacher, all respondents said no. Even if they taught mathe-

---

[2] At the time for the group interview Gunilla is teaching Swedish as a second language.

matics they did not recognise a mathematics teacher "in the mirror that they [held] up to themselves" (McNally et al., 2008, p.295). The mathematics teaching the respondents had done had not made them receive feedback, from themselves or others (Gee, 2000-2001), in line with being *a kind of mathematics teacher*. Not recognising themselves as a mathematics teacher can be connected to what it, according to the respondent, implies to be a mathematics teacher.

Before graduation the respondents express a clear opinion regarding how mathematics ought to be taught in primary school and they want to reform mathematics teaching. After graduation mathematics teaching has a concealed and limited role in the respondents' work. My focus in interviews and observations during the two years after graduation is mathematics but, in the work of the respondents mathematics has a concealed and limited role. Answering *no* to the question of being a mathematics teacher two years after graduation is a natural answer for some of the respondent for whom mathematics teaching is not included in their teacher assignment at that time. But nor those who work as class teachers, teaching mathematics every day, express having a sense of themselves as a mathematics teacher.

All respondent had chosen mathematics as one of their main subjects in teacher education. However, based on this study it does not seem to be enough to have a teaching degree including mathematics as a main subject to develop a sense of yourself as *a kind of mathematics teacher*. One aspect in developing a sense of yourself as *a kind of mathematics teacher* seems to be teaching mathematics, but that does not seem to be enough either. Gunilla has taught mathematics as a short run substitute teacher and in time for the group interview Nina is a class teacher, teaching mathematics in two different classes but this does not make them feel like mathematics teachers.

Gunilla's expression "I don't feel like a mathematics teacher but I can absolutely see myself having a job as a class teacher within which teaching mathematics is a part" indicated that there are different degrees of being a mathematics teacher. As mentioned, *the image of a (mathematics) teacher* is the similarities found in the unique experiences and expressions of the respondents regarding what it implies to be a (mathematics) teacher. Together the respondents express two different images of a teacher teaching mathematics (figure 1).

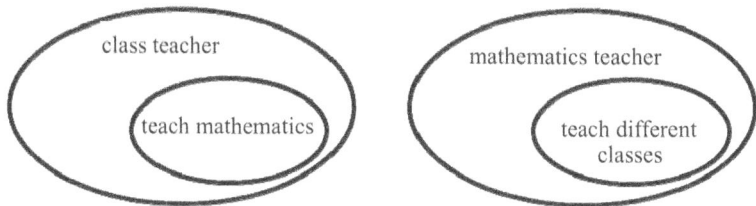

Figure 1　Two different images of teachers teaching mathematics.

The image to the left is in line with the teaching degree and the work of the respondents. However, that image is not, according to the respondents, a mathematics teacher. A mathematics teacher is the image to the right and none of the respondent identify with that. As such, their *image of a mathematics teacher* prevents them from recognising themselves as a mathematics teacher.

6　Conclusion and Discussion

As mentioned, to be recognised (by oneself and/or others) as *a kind of person* in a given context is a neverending process. After graduation mathematics teaching has a concealed and limited role in the respondents' professional identity development and above it was shown that their *image of a mathematics teacher* prevented them from recognising themselves as a mathematics teacher. According to George (2009) some student teachers are not given the opportunity to renegotiate their mathematical identity in teacher education and by that they bring psychic baggage from their own schooling into teaching. The respondents in this study had indeed, in teacher education, renegotiated their view of how mathematics should best be taught. But, what about their view of a mathematics teacher and its connection to this "new" way to teach mathematics?

According to Lindström Nilsson (2012), student teachers often retain the image they have of teachers when they start teacher education throughout their whole teacher education. According to van Bommel (2012) student teachers who are to become primary school mathematics teachers need to shift from seeing themselves as general teachers to locking at themselves as mathematics teachers. Van Bommel studied primary school student teachers during a mathematics education course and the requested shift was not made by the student teachers and neither was it addressed by the teacher educators. This is similar to other countries were most of the preparation primary school teachers receive places low emphasis on mathematics content in relation to the overall program which results in mathematics teaching never being put in the front in professional identity development (Tatto, Lerman & Novotná, 2009).

The results in the here presented study indicate that mathematics teacher is not a part of the professional primary school teacher identity of the respondents. But, when you, as a primary school teacher, are teaching mathematics you are a mathematics teacher. According to Palmer (2010), establishing a professional identity is about "picking up" the codes and the language associated with that profession. The codes and the language associated with the primary school teaching profession are seldom connected to mathematics but, instead, to caring and motherhood. Since the beginning of the 1900s, teachers teaching younger children have been pictured as warm, protecting and responsible females, a picture that still remains in both politics and the media.

To become recognised, by oneself and others, as a mathematics teacher, an individual's patterns of participation have to be in line with their image of the patterns of participation of a mathematics teacher (Palmér, 2013). With the image of mathematics teachers expressed by the respondents in this study it will become hard for them to recognise themselves as a mathematics teacher. For the respondents to develop (and striving towards developing) a sense of themselves as *a kind of mathematics teacher*; mathematics ought to become a part of their primary school teacher identities. Mathematics ought to become a part of their *image of a primary school teacher* as an *image of a primary school mathematics teacher.* Maybe then, mathematics teaching will become something they emphasise in their work.

According to Hodgen and Askew (2007) it is possible for primary school teachers to develop an identity as a teacher of mathematics but for this to happen the teacher has to "reconnect with mathematics whilst maintaining an identity as a primary teacher" (p.482). Just increasing the amount of mathematics courses in teacher education does not seem to be the solution (content knowledge was emphasised in the respondents' teacher education at the expense of decreased practice periods) but to connect mathematics to the student teachers' image of a primary school teacher. Then, maybe they will make a shift from looking at themselves as "only" general teachers to locking at themselves as also being mathematics teachers emphasising mathematics teaching in their work.

7 References

Aspers, P. (2007). Etnografiska metoder. Malmö: Liber AB.

Beijaard, D., Meijer, P.C. & Verloop, N. (2004). Reconsidering Research on Teachers' Professional Identity. Teaching and Teacher Education, 20(2), 107-128.

Bjuland, R., Luiza Cestari, M. & Borgersen, H.E. (2012). Professional Mathematics Teacher Identity: Analysis of Reflective Narratives from Discourses and Activities. Journal of Mathematics Teacher Education, 15(5), 405-424.

Borko, H. (2004). Professional Development and Teacher Learning: Mapping the Terrain. Educational Researcher, 33(8), 3-15.

Charmaz, K. (2006). Constructing Grounded Theory. A Practical Guide through Qualitative Analysis. London: SAGE Publications Ltd.

Gee, J.P. (2000-2001). Identity as an analytic lens for research in education. Rewiew of Research in Education, 25, 99-125.

George, P. (2009). Identity in Mathematics. Perspectives on Identity, Relationships, and Participation. In L. Black, H. Mendick & Y. Solomon. (Eds.), Mathematical Relationships in Education. Identities and Participation (pp.201-212). New York: Routledge.

Hammersley, M. & Atkinson, P. (2007). Ethnography. Principles in Practice. Third edition. New York: Routledge.

Hogden, J. & Askew, M. (2007). Emotion, Identity and Teacher Learning: Becoming a Primary Mathematics Teacher. Oxford Review of Education, 33(4), 469-487.

Kelle, U. (2007). The development of cathegories: Different approaches in Grounded Theory. In A. Bryant & K. Charmaz (Eds.), The SAGE Handbook of Grounded Theory. Los Angeles: SAGE Publications

Lerman, S. (2000). The Social Turn in Mathematics Education Research. In J. Boaler (Ed.), Multiple Perspectives on Mathematics Teaching & Learning (pp.19-44). Westport, CT, USA: Greenwood Publishing Group.

Lindström Nilsson, M. (2012). Att bli lärare i ett nytt skollandskap. Identitet och utbildning. Riga: Arkiv Förlag

McNally, J., Blake, A., Corbin, B. & Gray, P. (2008). Finding an identity and meeting a standard: Connecting the conflicting in teacher induction. Journal of Education Policy. 23(3), 287-298.

Palmer, A. (2010). Att bli matematisk. Matematisk subjektivitet och genus i lärarutbildningen för de yngre åldrarna. Stockholm: Stockholms Universitet.

Palmér, H. (2013) To become – or not to become – a primary school mathematics teacher. A study of novice teachers' professional identity development. Linnaeus University Dissertations: Linnaeus University Press

Peressini, D., Borko, H., Romagnano, L., Knuth, E. & Willis, C. A (2004). Conceptual Framework for Learning to Teach Secondary Mathematics: A Situative Perspective. Educational Studies in Mathematics, 56(1), 67-96.

Ponte, J.P. & Chapman, O. (2008). Preservice Mathematics Teachers' Knowledge and Development. In L.D. English, M.B Bussi, G.A. Jones, R.A. Lesh, B. Sriraman & D. Tirosh (Eds.), Handbook of International Research in Mathematics Education (pp.223-261). London: Routledge.

Sachs, J. (2001). Teacher Professional Identity: Competing Discourses, Competing Outcomes. Journal of Education Policy, 16(2), 149-161.

Schifter, D. (1996). Whats's Happening in Math Class? Envisioning new Practices through Teacher Narratives. New York: Teachers College Press.

Skott, J. (2010). Shifting the Direction of Belief Research: From Beliefs to Patterns of Participation. In M.F. Pinto & T.F. Kawasaki (Eds.), Proceedings of the 34th Conference of the International Group for the Psychology of Mathematics Education, 4 (pp.193-200). Belo Horizonte, Brazil: PME

Skott, J., Moeskær Larsen, D. & Hellsten Østergaard, C. (2011). From beliefs to patterns of participation – shifting the research perspective on teachers. Nordic Studies in Mathematics Education, 16(1-2), 29-55.

Tatto, M.T., Lerman, S. & Novotná, J. (2009). Overview of Teacher Education Systems Across the World. In R. Even & D.L. Ball (Eds.), The Professional Education and Development of Teachers of Mathematics. The 15th ICMI Study (pp.15-23). NewYork: Springer

van Bommel, J. (2012) Improving Teaching, Improving Learning, Improving as a Teacher. Mathematical Knowledge for Teaching as an Object of Learning. Faculty of Technology and Science. Karlstad: Karlstad University Studies.

Wenger, E. (1998). Communities of Practice. Learning, Meaning, and Identity. Cambridge: Cambridge University Press.

The Use of Technology in Calculus Classrooms – Beliefs of High School Teachers

Ralf Erens, Andreas Eichler
University of Education Freiburg | University of Kassel

Content

1 Introduction 134
2 Theoretical Framework 135
3 Method... 136
4 Results 138
    4.1   Instrumental initiation 138
    4.2   Instrumental exploration 138
    4.3   Instrumental reinforcement 139
    4.4   Instrumental symbiosis.. 140
    4.5   Teachers' beliefs across the levels of instrumental integration 141
5 Discussion and Conclusion... 142
6 References 143

Abstract

This report focuses on a part of a research programme concerning secondary teachers´ beliefs towards their teaching of calculus with particular attention to graphing and computer-algebra technology. First the theoretical framework and methodology is outlined. Afterwards the focus lies on studying these teachers´ beliefs with a particular concern to the teachers´ intended calculus teaching in technology-based secondary mathematics courses and the way teachers actually employ the technological device in teaching and learning of calculus. We conclude the paper by summarising our main findings and making suggestions for further research.

1 Introduction

For mathematics classrooms it has been well established that teachers' beliefs are a decisive factor for the teachers decisions what mathematical content is appropriate for a classroom (content), why this content is appropriate (teaching goals), and how the content should be taught (Calderhead, 1996). Thus, the main aim of our research programme is to investigate (upper secondary) teachers' beliefs for understanding their "professional world" (Calderhead, 1996, p. 709).

Indeed, in the recent decades research yielded evidence that teachers' beliefs strongly determine their intended teaching consisting of responses to the what, the why and the how (e.g. Schraw & Olafson, 2002; Philipp, 2007). Further, teachers' beliefs seem to impact on the teachers' classroom practice and also students' learning (e.g. Staub & Stern, 2002). However, the relationships between the teachers' espoused beliefs and both the classroom practice and the students' learning are not completely investigated (e.g. Philipp, 2007; Skott, 2009). Amongst others, there are two characteristics of teachers' beliefs that seem to influence the relevance of teachers' espoused beliefs for their enactment in classrooms, i.e. the centrality of teachers beliefs and the specificity of these beliefs referring to a mathematical domain (Franke et al., 2007; Eichler, 2011; Schoenfeld, 2011).

Referring to the aim to understand teachers' professional worlds, the focus of our research is on mathematics teachers' beliefs that are relevant for both the teachers' instructional planning (intended curricula), and the teachers' classroom practice. For this reason, we concern the specificity of upper secondary teachers' beliefs by investigating these teachers with a specific focus on a mathematical domain, which is, in this report, calculus. Recent steps in this research concern the identification of teachers' central beliefs towards calculus teaching that we reported elsewhere (Eichler & Erens, transmitted). Referring to these central beliefs, a striking result was attained by the considerable differences of teachers' beliefs towards using technology in classrooms. In this report, we focus primarily on this aspect, i.e. teachers' beliefs towards using technology in classrooms. Doing this, we firstly outline the theoretical framework referring to the use of technology in mathematics classrooms, and the construct of teachers' beliefs. Afterwards we discuss the method of our research. Finally we focus on results concerning calculus teachers' beliefs referring to different levels of integration of technology in classrooms.

2 Theoretical Framework

Research has ascertained that the role of technology in mathematics teaching requires an assiduous distinction between technical and conceptual mathematical activity. Technical activity is primarily concerned with tasks of procedural performance, whereas conceptual activity is concerned with tasks of inquiry, conjectures and justification (Zbiek et al, 2007). Thus, a central question for the teaching practice in technology supported classrooms is the function of technology in learning mathematics effectively. Although the function is distinguished further (Zbiek, 2007), we primarily focus on the activities teachers intend to enact in their classrooms.

The students' development from technical activities to conceptual activities is deeply described in the construct of instrumental genesis. Central to this theory is the notion of an "instrument", which is differentiated from an "artefact". The notion of instrument is a psychological one and not a description of a material artefact (Zbiek et al., 2007). The artefact (e.g. a calculator) and its capacity requires to be understood by the user who develops a relationship with the artefact. Thus, instrumental genesis is the process of the artefact (here: calculators) becoming an instrument and specifically how the artefact becomes a mathematical instrument – a tool that the user can employ for mathematical purposes.

Main ideas of the approach of instrumental genesis conducted by teachers are described by Goos et al. (2011). Providing an example relevant for calculus, the construction of a secant or tangent can be seen as an instrument to conceptualize the difference quotient. The students must learn to construct the secant and drag it along the graph of the function up to a given point (instrumentalisation). But they also have to learn, why dragging the secant is meaningful (instrumentation) and that this process leads to the conceptualization of a new mathematical definition. Goos et al. (2011, p. 313) outline that "instrumental integration is a means to describe how the teacher organizes the conditions for instrumental genesis of the technology proposed to the students and to what extent (s)he fosters mathematics learning through instrumental genesis" distinguishing four different levels of instrumental integration:

- Instrumental initiation: The focus lies on making the students familiar with the basic technical aspects of the tool by way of given tasks that enables students to use the technology for mathematical activity. Sometimes this aspect is also referred to as "tool competence".

- Instrumental exploration: The emphasis lies on students' exploration of the different features the technology offers through mathematical tasks. The teachers' aim may consist of improving the tool competence and deepen some mathematical knowledge.

- Instrumental reinforcement: On this level the students use the technology when they are faced with difficulties (e.g. solving an equation) while working on a mathematical task. The technology helps the students to overcome the difficulties so that they can concentrate on the mathematical knowledge.
- Instrumental symbiosis: The focal point on this level is that the technology is used by students to solve mathematical tasks with the explicit assistance of the technology so that instrumental integration is necessary to create mathematical content.

The question of how a "teacher organizes the conditions for instrumental genesis" (ibid.) is closely connected to the teachers' beliefs about what parts of an instrumental genesis facilitate their students learning. Defining beliefs as an individual's personal conviction concerning a specific subject, which shapes an individual's way of both receiving information about a subject and acting in a specific situation (Pajares, 1992), we understand the teachers' teaching goals referring to the instrumental genesis in technology supported classrooms as a specific form of beliefs. Since a teacher potentially has various specifications to the what (content), the why (goals) or the how (ways of teaching), we use the construct of belief systems (Green, 1971; Thompson, 1992) to describe the teachers decisions about the what, the why and the how, each representing teaching goals of different range of influence. In this report, we do not emphasise the identification of central beliefs that is a crucial aspect of belief systems (Thompson, 1992) since we highlighted this aspect elsewhere (Eichler & Erens, transmitted). Thus, we postulate the centrality of the teachers' beliefs outlined in the following section referring to technology supported classrooms, but analyse these central beliefs concerning the four levels of instrumental integration mentioned above.

3 Method

The sample for this study consists of 30 calculus teachers divided into three subsamples: pre-service teachers, teacher trainees and experienced teachers.

The first subsample includes 10 experienced teachers who have been teaching calculus for at least five years. Data concerning the intended curricula of these teachers that are assumed to be relatively stable (Calderhead, 1996) were collected once (table 1). The other subsamples consist of each 10 prospective teachers and 10 teacher trainees. The data for these subsamples were collected twice within one and a half years (table 1), since we assume that prospective teachers' beliefs potentially change when the teachers get their first intense practical experience.

Table 1: Data collection

Teacher sub-sample	End of university studies	Middle term of teacher training	Beginning of career as qualified teacher	After 5 years of teaching experience
Pre-service teachers	x	x		
Teacher trainees		x	x	
Experienced teachers				x

The teachers who participated in this study were recruited from different universities, teacher training colleges and schools across the southwestern part of Germany. However, our sample is a theoretical sample (Glaser & Stauss, 1967), but not a representative sample.

We used semi-structured interviews for data collection. Topics of these interviews were several clusters of questions that concern the content of calculus teaching, the related goals, reflections on the nature of calculus as a discipline generally, on the possible influence of technology on the students' learning, or textbook(s) used by the teachers. Further, we use prompts to provoke teachers' beliefs, e.g. potential challenges implied by the use of technology, fictive or real statements of teachers concerning instructional objectives and the use of technology in their calculus teaching.

For analysing the data, we used a qualitative coding method (Kuckartz, 2012) that is close to grounded theory (Glaser & Strauss, 1967). The codes gained by interpretation of each episode of the verbatim transcribed interviews indicate goals of calculus teaching. We used deductive codes derived from a theoretical perspective (cf. Grigutsch et al., 1998) and inductive codes for those goals we did not deduce from existing research concerning calculus education (Kuckartz, 2012). In this paper we primarily focus on the (inductive) codings relevant to the introduction and use of technological devices in the classroom not neglecting, however, the overall reconstruction of teachers´ belief system. The codings were conducted by at least two persons and we proved the interrater reliability to show an appropriate value.

4 Results

In this section, we discuss empirical evidences for the existence of the four levels of instrumental integration introduced above in the intended curricula of calculus teachers. Our aim here is to distinguish teachers' goals and conceptions how and why they use a graphing or a symbolic calculator in their calculus classrooms.

4.1 Instrumental initiation

Referring to the level of instrumental initiation there is some evidence that it is natural for (especially pre-service) teachers to realize that their commitment to using technology is similar to their students' needs:

> Mr. A: In my university courses I worked a little bit with tools that can produce graphs. At school I will have to become acquainted with these graphing handhelds quickly so that I will be able to understand students, who don't know how to use the calculator and help them to solve these problems.

Primarily pre-service teachers and teacher trainees report in the interviews that they are prepared to meet the challenges they face on a basic instrumental level.

As all the teachers of our sample use technology in their classroom following a compulsory curricular requirement to use a graphing handheld or a computer algebra calculator, it is not surprising that many of the pre-service and trainee teachers consider themselves to be on the level of instrumental initiation. The consequences for implementing learning sequences how to use the technology in their enacted curriculum is thus natural and need not be elaborated further.

4.2 Instrumental exploration

Although (pre-service) teachers after their university studies do not have much teaching experience they affirm their view that the use of technology will make their calculus teaching more dynamic and effective as compared to a chalk-and-blackboard approach. These teachers mention that the appropriate use of technology enables the visualization of mathematical concepts and provides opportunities to utilize multiple representations to further students' comprehension of more abstract concepts:

> Mr. B: "The surplus of using technology I clearly see that we can very easily produce graphs and thus visualize particular properties of functions.
>
> Mr. C: "Changes due to technology use? Well, I notice that my students deal with tables of real data more audaciously. In former times they had numerous difficulties

to transfer these to a graph and so on. With graphing calculators they can put the data in, graph it and then think about the functional properties."

Both examples (teacher trainee & experienced teacher) show the modes of integrating technology in order to assist students' learning of calculus with respect to two aspects: nurturing students' comprehension of functions so that they can form a mental picture as well as fostering the relationship between the tool and the mathematical knowledge. The benefit of changing between different representations (table, graph & function) of mathematical objects is noteworthy here. Mr C is a good example to show that students' activities are enhanced by the use of technology on a technical as well as a mathematical level. In our sample there are more teachers who mention the aspect of visualization than those who emphasize the possibility of being able to change between three modes of representation. Regarding the degree of professionalization, teachers of all three subsamples designate visualization as the main additional benefit of using technology. After instrumental initiation of how to use the specific ability of the artefact (i.e. producing graphs) the teachers' objective of visualizing functions produces the desired links with the mathematical knowledge. Referring to the different levels of instrumental integration the emphasis of utilizing the features offered by the technology to broaden the mathematical knowledge can thus be seen as instrumental exploration.

Regarding the level of instrumental exploration, Thomas (2006) reports that teachers consider "the time and effort needed by both students and teachers in order to become familiar with the technology" to be problematic. Although this aspect is at most peripherally mentioned in our data, nearly all teachers in our sample mention that the possibilities offered by the technology is of greater importance. Moreover, many teachers specify that compared to calculus lessons without digital technology there occurs a saving of time once the students are familiar with the artefact.

> Mrs D: "In former times teachers had to copy students' solutions onto an overhead transparency. Using the handheld displays practically saves a lot of time."

4.3 Instrumental reinforcement

Most of the teachers who mentioned the introduction of handhelds as a gain of teaching time express clear intended purposes for this gained time:

> Mrs E: "Often my students use the calculator in order to check their results. First they think about their strategy of solving the given task and then they can reassure themselves about solutions."
>
> Mrs F: "Of course the students do not have to do so many time-intensive calculation routines any more. That's a great relief for many of my students. We can use that to

concentrate more on the subject matter of these calculus concepts. There is a shift to analyzing and interpreting results rather than these schematic calculations."

Many teachers express beliefs in a similar way representing a central characteristic of instrumental reinforcement, i.e. the benefit of using technology to overcome obstacles when solving problems or to concentrate on the mathematical knowledge. The quoted teachers mention the function of checking results and relieving students from procedural investment as main benefit of technology supported classrooms. In particular, for many teachers relieving students from procedural investment is related to creating a positive emotional attitude in connection with their calculus courses. In teachers' mathematical views calculus und tedious calculations (derivates etc.) are often seen as an obstacle for successful student learning. In making the concession that repetitive schemes need not be enacted extensively, teachers see an opportunity to further their students' comprehension in the direction of a more advanced mathematical thinking (Tall, 2008). Further, allowing students to overcome algebraic difficulties with the artefact as a manipulating and calculation assistant, many teachers believe that the tool is a meaningful device for students learning and a means to concentrate on the conceptual aspects of calculus at school.

4.4   Instrumental symbiosis

As a paradigmatic example of a teacher with a high level of instrumental integration representing the level of instrumental symbiosis, we use the case of Mrs H. As an experienced teacher she uses a computer-algebra calculator (CAS) which includes the opportunity of symbolic manipulations (in addition to tables & graphs):

> Mrs H: "With CAS handhelds I can make students experiment themselves e.g. finding out about characteristics of several functions and relevant structures. Formerly [i.e. without technology use] I used to introduce that in more teacher-centred way using only one function as you had to draw a graph which was rather time-consuming."

Mrs H emphasises the benefit of technology use to achieve a generic development of calculus concepts. Moreover a contrast to her former teaching practice is mentioned regarding several elements: the shift from an instructivist to a more constructivist and inquiry-oriented teaching orientation. This approach facilitates students' conceptual learning in experimenting with mathematical objects in an inductive way:

> Mrs. H: "The CAS is simply a tool that provides the opportunity of checking certain mathematical things quickly and my students can actually see whether a conjecture that they have found out themselves with one function can be validated with other types of functions."

Mrs. H mentions that making students true participants in mathematical learning is an essential part of her calculus teaching. The above mentioned opportunities of a meaningful use of technology further support her overarching teaching objective of reaching a higher standard of abstraction:

> Mrs. H: "If they (i.e. students) have affirmed their hypotheses with 5 examples, they are often willing to get involved in actually verifying the result in a more formal way, because they really want their own results to be proven which they have thought of."

Consistently describing the intense integration of technology into her calculus course, Mrs. H describes mainly three aspects of technology as advantages for teaching calculus, i.e. the benefit of technology of exploring mathematical phenomena, of conceptualisation, and of checking results or hypotheses. Students in Mrs. H's courses are faced with mathematical tasks that allow them to utilize their technical knowledge of the artefact and connect this to new mathematical knowledge.

4.5 Teachers' beliefs across the levels of instrumental integration

Across all levels of instrumental integration, particularly experienced teachers also consider disadvantages referring to technology supported classrooms, e.g. a probable decline in students' competencies to perform relevant calculation procedures that are fundamental for calculus teaching:

> Mr. K: "There seems to be a development that computers are pervasive in every discipline and hardly anyone works with just a paper and pencil. Some competencies go missing (…) as I said there are advantages but also disadvantages using calculators."

For example, Mrs. L meets the mentioned disadvantage by emphasising both technology use, and paper and pencil skills:

> Mrs. L: "[using technology] some things are quicker. But on the other hand, students must have the proficiency for both – doing these calculations by hand AND have the know-how to do with their calculator – the time factor must not be neglected…"

These experienced teachers emphasise the benefit of technology use. However, they tend to hold beliefs pointing out that they put considerable effort in exercises and calculation of routine tasks having in mind the students' preparation of final exams, which consist of a technology-free part.

Finally, one teacher expresses further beliefs that describe an obstacle of using technology in calculus courses:

> Mr. M: "They [i.e. students] use the calculator and take results as an absolute truth without reflecting on them. Does it really make sense? Taking into account that e.g. one might have typed wrong numbers is unimaginable."

For Mr. M the integration of technology in calculus courses poses another challenge for the teacher role as a mediator of learning processes. Students may be tempted to regard the technological device as a "universal remedy" for mathematical problem-solving. The mathematical fidelity (Zbiek et al., 2007) of the tool that students take for granted may lead to disparities in technology-based learning and teaching. One possible pitfall, for example, is the difference between graphic display and the interpretation of the display when considering the sine function with small periods. Further examples can be found in the relevant literature (e.g. Zbiek et al., 2007).

5 Discussion and Conclusion

In approaches using technology in mathematics teaching, the teacher plays a significant role in how the technology is used in learning (Tall, 2008). Moreover, teachers are experts of their teaching and teachers´ conceptions and beliefs are a decisive factor for the what, the how and the why. Given the circumstances of the mathematics syllabus and the technological environment, teachers may be persuaded of the value of technology integration in their actual teaching or not. In reconstructing teachers´ beliefs related to technology the above results provide a basis for understanding teachers´ objectives and goals in the framework of instrumental genesis.

Regarding our sample of teachers, it seems possible to ascertain two antithetical belief systems referring to the integration of technology in upper secondary calculus courses, i.e. beliefs systems that we call "the old school" and "technology supporter".

Teachers describing themselves as belonging to "the old school" consistently express severe doubts about technology use in helping students´ mathematical learning that we exemplarily quoted in the last paragraph. These resistances cannot be attributed to professional problems on the instrumental level. Rather personal and epistemological factors account for these deeply-rooted objections in their belief system. In addition to the quoted doubt referring to technology supported classrooms, some teachers emphasize a deductive structure of content as their main teaching objective and underline that it is necessary for students to fully understand mathematical ideas before using any kind of technological help.

In contrast to "old-school" teachers there are "technology supporters", i.e. teachers, who expressly utilize graphic and symbolic technology in order to realize a problem-oriented approach to teaching calculus. The benefit of technology is seen in various aspects of their teaching: making students true participants in learning calculus (exploring phenomena themselves) with a shift from an instructivist to a more inquiry –oriented way of teaching as well as the advantages for conceptualisation or checking results by using the artefact.

In between our data suggests that teachers' beliefs towards technology range in a continuum that reaches from scepticism to reluctance up to those teachers who use the given technology implemented in the curriculum as a means to an end. This end is specified with single but often mixed factors: making concessions to students (e.g. no tedious procedures) and gaining time in order to concentrate on modelling tasks or being able to grant more classroom time to qualitative and heuristic work and to initiate an advanced mathematical thinking.

All teachers of our sample express that regardless of their degree of professional development or their individual conceptions about technology that paper and pencil-skills are more or less important. Particularly for the experienced teachers these skills are mandatory in terms of students' needs referring to final exams but also as a crucial part of mathematical thinking.

To address the challenge of a meaningful integration of technology into calculus teaching at upper secondary level, further research needs to be carried out in several directions. In the context of our research programme one direction concerns gaining evidence towards the relevance of teachers' (technology) beliefs for their actual teaching. The other direction includes the relationship between teachers' beliefs and students' learning. As changes in mathematical learning in the digital age are difficult to implement in school, more research is needed by expounding the relationships between cognition, affect and their impact on teachers and students in mathematics classrooms.

6 References

Calderhead, J. (1996). Teachers: Beliefs and knowledge. In D. C. Berliner (Ed.), *Handbook of Education* (pp. 709-725). New York: MacMillan.

Eichler, A. (2011). Statistics teachers and classroom practices. In C. Batanero, G. Burril, & C. Reading (Eds.), *Teaching Statistics in School Mathematics-Challenges for Teaching and Teacher Education* (pp. 175-186). New ICMI Study Series, vol. 15. Heidelberg, New York: Springer.

Eichler, A. & Erens, R. (transmitted). Teachers' beliefs towards teaching calculus. Transmitted for *ZDM – The International Journal on Mathematics Education* 46(4).

Franke, M. L., Kazemi, E., & Battey, D. (2007). Understanding teaching and classroom practice in mathematics. F. K. Lester (Ed.), *Second handbook of research on mathematics teaching and learning* (pp. 225-256). Charlotte, NC: Information Age Publishing.

Glaser, B. & Strauss, A. (1967). *The discovery of grounded theory.* Chicago: Aldine.

Goos, M. & Soury-Lavergne, S. (2010). *Teachers and Teaching: Theoretical Perspectives and Issues Concerning Classroom Implementation*. In Hoyles, Celia (ed.) et al., Mathematics education and technology-rethinking the terrain. The 17th ICMI study. Dordrecht: Springer

Green, T. (1971). *The Activities of Teaching*. New York, NY: McGraw-Hill.

Grigutsch, S., Raatz, U., & Törner, G. (1998). Einstellungen gegenüber Mathematik bei Mathematiklehrern. *Journal für Mathematikdidaktik 19* (1), 3-45.

Kuckartz, U. (2012). *Qualitative Inhaltsanalyse. Methoden, Praxis, Computerunterstützung*. Weinheim/Basel: Beltz Juventa.

Pajares, M. F. (1992). Teachers' beliefs and educational research: Cleaning up a messy construct. *Review of Educational Research*, 62(3), 307-332.

Philipp, R.A. (2007). Mathematics teachers' beliefs and affect. In F. K. Lester (Ed.), *Second handbook of research on mathematics teaching and learning* (pp. 257-315). Charlotte, NC: Information Age Publishing.

Schoenfeld, A. (2011). *How we think—a theory of goal-oriented decision making and its educational applications*. New York: Routledge.

Schraw. G. & Olafson, L. (2002). Teachers' epistemological world views and education practices. *Issues in Education: Contributions from Educational Psychology*, 8 (2), 99-148.

Skott, J.(2009). Contextualising the notion of 'belief enactment'. *Journal for Mathematics Teacher Education* 12, 27-46.

Staub, F., & Stern, E. (2002). The nature of teacher's pedagogical content beliefs matters for students' achievement gains: quasi-experimental evidence from elementary mathematics. *Journal of Educational Psychology, 94* (2), 344-355.

Tall, D. (1991). The Psychology of Advanced Mathematical Thinking. In D. Tall (Ed.), Advanced mathematical thinking (pp. 3–21). Dordrecht, The Netherlands: Kluwer.

Tall, D., Smith, D. & Piez, Cynthia (2008). Technology and calculus. In M.K. Heid & G.W. Blume (Eds.), *Research on technology and the teaching and learning of mathematics: synthesis, cases and perspectives. Vol. 1: Research synthesis* (207-258). Charlotte, NC: Information Age Publishing.

Thomas, M.O.J. (2006). Teachers using computers in mathematics: a longitudinal study. In C. Hoyles, J. Lagrange, L.H. Son & N. Sinclair (Eds.), *Proceedings for the Seventeenth ICMI Study Conference: Technology Revisited, Hanoi University of Technology, 3rd–8th December, 2006* [CD-ROM].

Thompson, A. G. (1992). Teachers' beliefs and conceptions: a synthesis of the research. In D. A. Grouws (Ed.), *Handbook of research on mathematics teaching and learning* (pp. 127–146). New Jersey: NCTM.

Zbiek, R M.., Heid, M. K., Blume, G., & Dick, T.P. (2007). Research on technology in mathematics education: The perspective of constructs. In F. K. Lester (Ed.), *Second Handbook of Research in Mathematics Teaching and Learning* (pp. 1169-1207). Charlotte, NC: Information Age Publishing.

Mathematics Student Teachers' Metaphors for Technology in Teaching Mathematics

Päivi Portaankorva-Koivisto
University of Helsinki

Content

1 Introduction 146
2 Teachers Beliefs about Using Technology in Mathematics Teaching 147
3 How to Categorize Teachers' Views of Using Technology in their Mathematics Teaching? 148
4 About Metaphors and Metaphor Theory 149
5 The Study 150
    5.1   Educational setting 150
    5.2   Method 151
6 Analysis and Results 152
7 Discussion 154
8 References 155

Abstract

While the metaphors for mathematics and for teaching mathematics have received some attention, there is a lack of research on metaphors for technology. This study sought to investigate mathematics student teachers' metaphors for technology in teaching mathematics. Based on metaphor theory and two theories of technology, the author analyzed 60 student teachers' metaphors for technology. The findings reveal that student teachers' views of using technology in mathematics teaching are ambiguous. The instrumental view of technology was dominating the data. Although the participating student teachers seem mainly to have a positive attitude towards technology, they need adequate opportunities in teacher education to explore the pedagogical and educational use of technology.

1 Introduction

Technology and computer-aided learning materials are nowadays an essential part of modern mathematics teaching. Technology can help the teacher to visualize mathematical concepts and give immediate feedback for students. Moreover, computer-based tools can help students to manipulate mathematical graphs and figures, and execute calculations that either cannot be done manually, or are too slow to calculate by hand. Therefore, knowledge of technology and computer-based resources is important for a future mathematics teacher. (Asikainen, Pehkonen & Hirvonen, 2013.)

Recently, the question of what mathematics teachers need to know in order to be able to integrate technology into their teaching has received much attention (see Mishra & Koehler, 2006; Akkoç, Bingolbali, & Ozmantar, 2008).

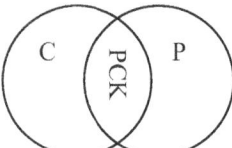

Figure 1   Shulman's pedagogical content knowledge (PCK) 1987

Pierson (2001) added a technology component to Shulman's (1987) PCK framework (figure 1) and described 'Technological Pedagogical Content Knowledge (TPCK)' as a combination of three types of knowledge: (1) content knowledge, (2) pedagogical knowledge, and (3) technological knowledge including the basic operational skills of technologies. Later illustrated as an intersection of three knowledge categories: technological, pedagogical and content by Mishra and Koehler (2006) (figure 2). Akkoç et al. (2008) propose, that the TPCK framework can guide teacher educators to design courses concerning technology as a part of mathematics teacher education programmes.

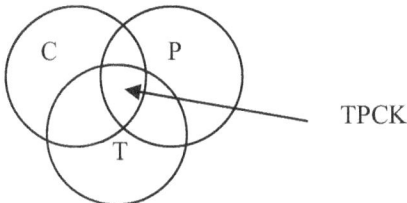

Figure 2   Mishra's and Koehler's technological pedagogical content knowledge (TPCK) 2006.

And further on, since many pre-service teachers might not have learnt their mathematical content with technology, they suggest that school mathematics should be revisited using various technological tools.

The teachers' characteristics play an important role in adapting technology in their teaching. According to Becta study (Becta, 2004), which reviewed the research literature on 'barriers to the uptake of ICT by teachers', a number of teacher-level barriers were identified. A very significant determinant of teachers' levels of engagement in ICT is their level of confidence in using the technology. Teachers with little or no confidence in using computers in their work will try to avoid them altogether. Levels of confidence are directly affected by the amount of personal access to ICT that a teacher has, the amount of technical support available, and the amount and quality of training available. Teachers are sometimes unable to make full use of technology because they lack the time needed to fully prepare and study materials for lessons. Besides technical faults with ICT equipment are likely to lead to lower levels of ICT use by teachers. Sometimes a total resistance can be seen in teachers' actions. Teachers are unwilling to change their teaching practices. Teachers who do not realise the advantages of using technology in their teaching are less likely to make use of ICT.

In Finnish context using technology in mathematics teaching polarized the teachers' responses. Some teachers thought that technology is important and can help students to learn. Others were more skeptical and emphasized the teacher's role in guiding. (Asikainen et al., 2013.)

2 Teachers Beliefs about Using Technology in Mathematics Teaching

Teachers' beliefs about mathematics, its learning and teaching are reflected strongly in the way they teach. Reflection is assumed to play a key role in change of practice and many researchers see a cyclical relationship between changing beliefs and changing practices. (Kagan, 1992; Lerman, 2002; Wilson & Cooney, 2002.) Already in the 90's Veen (1993) suggested that whether teachers use or not use computers depends on two basic factors: the school level and the teacher level. At the school level the principals are responsible for financial, organizational and moral support and they should provide a long-term perspective. Instead, at the teacher level the teachers adopt new media if they can use them in accordance with their existing beliefs and practices. By then the only major study to examine the relationships among teachers' epistemological beliefs, pedagogical beliefs, and their instructional uses of technology was conducted as a part of an evaluation of the Apple Classrooms of Tomorrow

(ACOT) project (Yocum, 1996) The study indicated that teachers tend to adopt new classroom practices based on whether the assumptions underlying the new practices are consistent with their personal epistemological beliefs (Yocum, 1996).

More recent studies suggest that there is a parallel between a teacher's student-centered beliefs about instruction and the nature of the teacher's technology-integrated experiences (Judson, 2006; Totter, Stutz, & Grote, 2006). In Becker and Ravitz's (2001) study, the results show that computer use among teachers is related to more constructivist views and practices and to changes in practice in a more constructivist-compatible direction. Burton (2003) also showed in her study with elementary teachers that this development can happen two-way. Professional development experiences involving technology will also facilitate a change in teachers' beliefs regarding teaching and learning towards a more student-centered focus. This may reflect the teachers' believes about their role. A traditional role will change to that of facilitator and partner in inquiry. Also Totter, Stutz and Grote (2006) suggest that teachers who adopt a student-oriented constructivist teaching style are more likely to make use of new technology in classrooms, and vice versa. They present some teachers' key characteristics, which influence the use of new media in classrooms. Positive influencing factors are openness to change, willingness to cooperate and constructivist teaching style. Moreover, negative influencing factors are lack of time, lack of ICT confidence and lack of ICT competence.

3 How to Categorize Teachers' Views of Using Technology in their Mathematics Teaching?

Chen (2011) provides two theories of technology as a framework for looking at the use of technology in mathematics teaching. According to the instrumental view of technology, the technology is seen as a tool or device assisting students to learn. It is an independent entity in this learning process. Teachers regarded technology as morally neutral and good in the class if used appropriately. Technology was also seen as empowerment while performing complex calculations, showing connections among different representations, and visualizing mathematical concepts. On the other hand according to the substantive view of technology, it represents an autonomous cultural system and acts to shape human perception and actions. Teachers in Chen's study (2011) regarded technology as inevitable and thought that it was their responsibility to equip their students with technology skills.

Gilbert and Kelly (2005) categorize attitudes towards technology pointing at emotions. Technology was seen either as a frontier, and a tool for exploration, or as a frontline, an attack and a defense. In White's (2004) study he proposes five teachers views of technology: a demon, a servant, an idol, a partner, and a liberator. When technology is seen as a demon, teachers actively oppose and decline to integrate technology into their teaching. Teachers feel either afraid or not willing to learn. When technology is seen as a servant, teachers assimilate technology in their teaching but into their existing instructional practices. If technology is seen as an idol, the emphasis is upon the teaching about computers rather than with computers. If technology is more like a partner, the students are actively engaged in working with data, and making meaning of their results. The technology has changed not only how students learn but also what they learn. When technology is seen as a liberator, technology is organizing and structuring the education itself.

In their study Levin and Wadmany (2006) provide a profound overview on technology and teacher change. Their findings reveal that following multi-year experiences in technology-based classrooms, teachers' educational beliefs changed quite substantively, yet revealing rather multiple views than pure beliefs. The categories used were (1) technical interest, (2) communicative or practical interests, and (3) emancipatory knowledge interest. A technical knowledge interest is being realized when technology is perceived as a means of practicing knowledge, skills, understanding, or competency, and when the context is not considered particularly relevant. When technology is serving a practical interest, then its role is in communication and interpretation. The emancipatory view of technology's role according to Levin and Wadmany (2006) involves the pursuit of knowledge or capacity to become conscious of the ways in which knowledge is constructed.

This study aims to find out the needs for support and encouragement the student teachers have in using technology in their mathematics teaching, and survey mathematics student teachers' views of technology by using metaphors. The research question is: What kind of views expressed with metaphors do the preservice mathematics teachers have of using technology in their teaching?

4 About Metaphors and Metaphor Theory

Metaphors are significant in teacher education. They provide insights into complex concepts of teaching and learning and thus provide a window into the comprehension of teachers' personal experiences. The word 'metaphor' has its roots in Greek and is based on word *metapherein*, meaning to transfer or carry. That means that something is carrying across, and thus by metaphor we denote that something is, in some sense, something that it literally is not. As metaphors focus on similarities, they can be used to express views of the nature of technology. While they provide a way of talking about current views of technology, metaphors can open up new ideas of thinking about these perceptions. (Lakoff & Núñez, 2000.)

Metaphors are not mere words or expressions. Instead they are ontological mappings across conceptual domains. "Mappings are not arbitrary, but grounded in the body and everyday experience and knowledge." (Lakoff, 1993, p. 245). Also a metaphor is not just a matter of language, but of thought and reason. "Metaphor is fundamentally conceptual, not linguistic, in nature." (Lakoff, 1993, p. 245). The challenge in using metaphors is the different knowledge and different experiences that people bring in while telling something via metaphors. Like Lakoff (1993, p. 245) proposes "metaphorical mappings vary in universality; some seem to be universal, other are widespread, and some seem to be culture specific", and "metaphor is mostly based on correspondences in our experiences, rather than on similarity".

While metaphors involve understanding one domain of experience in terms of a different domain of experience, metaphors can allow student teachers to understand and express abstract matters in concrete ways, and as Noyes (2006) points out that metaphors can reveal hidden beliefs of mathematics and help teacher educators to create conflict situations that might shift the meanings of mathematics. Reeder, Utley and Cassel (2009) argue that if experiences in teacher education programmes are to bring about meaningful transformation for preservice teachers, teacher educators must provide opportunities for students also to critically examine their own thinking and beliefs about teaching and learning.

5 The Study

This study was conducted among pre-service mathematics teachers during their didactical course in the beginning of 2013. In this chapter the secondary school teacher preparation programme in Finland is shortly introduced. After that the data gathering and instruments are brought forward.

5.1 Educational setting

In Finnish secondary school, teacher preparation is a 5-year programme (3 BA and 2 MA). The students major in one subject, and minor in one or two other subjects (e.g. mathematics major, and chemistry and physics minor). This means that the students take education as minor and these teacher's pedagogical studies (60 ECTS) can be taken within one academic year. Usually the students do their pedagogical studies at the end of their BA studies. The programme gives general teacher qualifications to teach children (7th grade, 12-13 years), young people (secondary school) and adults in educational institutions offering general, vocational and adult education. Moreover, according to programme objectives, the future teachers gain a starting point to develop into a professional who plans, implements, evaluates, and develops teaching. In pedagogical studies the student teachers have to combine content knowledge, knowledge related to education and different learners, pedagogical content knowledge (i.e. knowledge of how to teach, study and learn the subject), and knowledge about school practices into their own pedagogical practical theory.

5.2 Method

Data for this study was gathered from 60 mathematics student teachers in January 2013. By then the student teachers in this study had been able to familiarize themselves with nearly half of their pedagogical studies, namely the first mathematics methods course and student teachers' first practical classroom experiences. The assignment was: the student teachers were asked to determine a statement "technology in mathematics teaching is", and to continue with an explanation of why it is so. They were not asked to identify themselves in their texts, so the texts remained anonymous. Still, only the metaphors with students' permission to use as data were gathered for this study.

The analysis was made in two phases: firstly, mere inductively, categories driven by the data; secondly, based on selected theory and previous studies. At first, all metaphors were read through thoroughly and five categories were formed. The metaphors were placed in categories exclusively, and a short description of categories was developed. Only three metaphors could not be categorized, merely because they were not true metaphors; they were ether lacking the metaphor word or the explanation. At this stage of analysis metaphors were assessed independently by the author. The five categories were:

(1) Useless: the metaphor expressed reluctance to use technology in mathematics teaching

(2) Over-advertised: the metaphor reflected the usefulness of technology but at the same time downplayed its role in teaching

(3) Tentative: the metaphor stressed that technology requires tackling, or it requires competencies, and it was not fitting for every teacher

(4) Good servant but bad master: the metaphor introduced both good and poor features of using technology in mathematics teaching

(5) Technology believer: the metaphor praised the technology and its role as a savior.

After the categories were formed and the descriptions were written down, an assistant classifier was used. Once independent data analysis was completed the findings were compared for inconsistencies and worked collectively to reconcile some of the inconsistencies. After categorization, categories could be arranged in order in respect of emotional aspects; the one end of the axis was anxiety and the other was enthusiasm. In the study of Gilbert and Kelly (2005) these endpoints were called frontline and frontier (c.f. .White, 2004).

At the second phase of analysis the metaphors were categorized again, but this time the dimension was weather technology was seen as a tool or it has intrinsic value or value in itself. These categories were adopted from Chen (2011), who provides two theories of technology as a framework for looking at the use of technology in mathematics teaching: the instrumental view of technology and the substantive view of technology.

6 Analysis and Results

Some of the metaphors were rather clear-cut, but some were opened up to various possible interpretations. The categorization was exclusive, so each metaphor was counted as one. There were 60 metaphors in total at the beginning of analysis, and three of them were left out of analysis, because they were not true metaphors. After the first reading only the metaphors categorized the way that both classifiers could agree on were accepted as data (see table 1). This resulted that the number of metaphors decreased from 60 to 37 metaphors.

Table 1 Classifiers' categorizations and the data that was selected.

Category	Useless	Over-advertised	Tentative	Good servant but bad master	Technology believer	Total
Both classifiers agreed	3	5	8	13	8	37
Classifier 1	3	6	17	19	12	57
Classifier 2	11	9	12	15	10	57

Following examples explain the differences between classifiers. The metaphor "Technology in mathematics teaching is like an SLR camera. Basically, a good device, but only a very few know how to use it right" was one of those metaphors in between categories. Classifier 1 was stressing the part "a good device, but" and categorized the metaphor in category 3. Classifier 2 stressed the part "a very few know" and categorized it in category 2.

Sometimes differences appeared when the first classifier was looking at the explanation and the second was stressing the metaphor. For example "Technology in mathematics teaching is a spice. When used incorrectly it can ruin even good ingredients, but well used it will take the remaining food ingredients to a new level". Classifier 1 categorized it in category 4, because of the explanation, and classifier 2 in category 3 because of the metaphor word that was used.

After the first categorization the same metaphors were categorized again but this time according to the role of the technology, weather technology has an intrinsic use or only use as a tool. Only the metaphors both classifiers were agreeing were selected to this analysis and the number of metaphors declined from 37 to 27 (see table 2).

Table 2 Technology metaphors categorized according to the role of the technology.

Category	Useless	Over-advertised	Tentative	Good servant but bad master	Technology believer	Total
Intrinsic value	0	2	3	0	5	10
Tool	2	3	2	10	0	17
Total	2	5	5	10	5	27

Most of the metaphors (17/27) were describing technology as a tool. However, all the metaphors in category *good servant but bad master* were tool metaphors (10/27), and all metaphors in category *technology believer* were metaphors where technology was having an intrinsic value (5/27).

Only 2/27 metaphors were categorized in the category of *useless*. Both of them were expressing the role of the technology as a tool. "Technology in mathematics education is a magician's smoke. Magician's smoke prevents the audience to see exactly what is going on. Sometimes the use of technology, in particular the terrible, cumbersome CAS calculators, can come between the student and the content and the student does not understand what he is doing or what he is seeing, and his attention goes to finger at the device." In these metaphors the technology was only interfering with the learning and teaching of mathematics.

One of the metaphors in category *over-advertised* was: "Technology in mathematics teaching is a Swiss pocket knife. With it one is able to do anything, and still it is rarely used." Almost half of these metaphors (2/27) were stressing the role of the technology as tool. One of the metaphors (3/27/) in this category expressing the intrinsic value of technology was "Technology in mathematics teaching is offering unlimited possibilities. However, no one is able to fully take advantage of them. Technology has advanced so rapidly, that it seems that teachers do not have time to follow the development."

In category *tentative* where 5/27 metaphors were categorized, some of the metaphors were manifesting of how much work the technology requires the teacher to do. "Technology in mathematics teaching is like gardening. It requires time and dedication, in order to get the perfect result." This metaphor was expressing the intrinsic value of technology. Some metaphors in this category were expressing the uncertainty of technology. "Technology in mathematics teaching is a journey into the unknown, because you never know what can be found in front of you, or when the journey ends." Like the previous metaphor also this is stressing the intrinsic value of technology. The metaphors in this category were also expressing the know-how teachers' need, when they are planning to use technology in their teaching. "Technology in mathematics teaching is a flash drive. It is of no use if one cannot use it." This metaphor was also expressing the role of the technology as a tool.

The metaphors (10/27) in category *good servant but bad master* were metaphors where the role of the technology was seen as a tool. "Technology in mathematics teaching is a dishwasher. Nice device that saves time and effort, but is not necessary." "Technology in mathematics teaching is a good servant, but a bad master. It takes too easily the focus away of the subject being taught." "Technology in mathematics teaching is like electricity in summer cottage. Without it, it'll be fine, but yes, it's a little miserable and dreary in the long run."

The last category was *technology believer*. All 5/27 metaphors in this category were stressing the intrinsic value of technology. "Technology in mathematics teaching is like the child's first step. It must be taken in order to go ahead." "Technology in mathematics education is like getting additional senses. It is like getting more eyes and more ears to use."

7   Discussion

The results of this study indicate that mathematics student teachers views of using technology in their teaching are moderately positive. When only affective categories were taken into account, no less than 61.7 % of metaphors were positive or fairly positive. When the role of the technology as an intrinsic value or as a tool was determinative 74.1 % of metaphors were positive or fairly positive. Weather these student teachers will use technology in their teaching in the future is still uncertain. Like in Becta study (2004) and also in the study of Totter,

Stutz and Grote (2006) negative influencing factors are lack of time, lack of ICT confidence and lack of ICT competence. Even 63.0 % of the metaphors were describing technology in mathematics teaching as a tool. This was also the case in Chen's (2011) study. Hopefully, like in Levin's and Wadmany's (2006, p. 174) three-year study the teachers who could integrate technology in their teaching to long-term basis developed their "viewing technology as a technical tool to seeing it as a partner that can empower the student, teachers and the learning environment".

The present study is significant and relevant for several reasons. First, it offers an important contribution to the exploration of teachers' professional growth when integrating technology component into student teachers' reflective practice. Second, it gives us teacher educators a view of mathematics student teachers' beliefs of technology and its usefulness in mathematics education. This study also revealed how difficult is to categorize metaphors. The person writing the metaphor may have totally different view on the chosen metaphoric expression. For example the flash drive, if one uses it all the time, it is a necessity and one might be surprised if someone else is not able to use it. For somebody else it would be a device not so often used, and almost useless. Depending on classifiers own interests and his cultural background the words get different meanings. This is why the metaphors are so intriguing and the metaphor theory continues to interest researchers.

8 References

Akkoç, H., Bingolbali, E., & Ozmantar, F. (2008). Investigating the Technological Pedagogical Content Knowledge: A case of derivative at a point. In *Proceedings of the Joint Meeting of the 32nd Conference of the International Group for the Psychology of Mathematics Education, and the XXX North American Chapter* (Vol. 2, pp. 17-24).

Asikainen, M. A., Pehkonen, E., & Hirvonen, P. E. (2013). Finnish Mentor Mathematics Teachers' Views of the Teacher Knowledge Required For Teaching Mathematics. *Higher Education Studies*, 3(1), p79.

Becker, H. J., Ravitz, J. L., & Wong, Y. (1999). Teacher and Teacher-Directed Student Use of Computers and Software. Teaching, Learning, and Computing: 1998 National Survey. Report# 3.

BECTA (2004) A review of the research literature on barriers to the uptake of ICT by teachers. British Educational Communications and Technology Agency.

Burton, D. B. (2003). Technology professional development: a case study. *Academic Exchange Quarterly*, 7(2), 2378-2381.

Chen, R. J. (2011). Preservice Mathematics Teachers' Ambiguous Views of Technology. *School Science and Mathematics*, *111*(2), 56-67.

Gilbert, J., & Kelly, R. (2005). Frontiers and frontlines: metaphors describing lecturers' attitudes to ICT adoption. *Educational Technology & Society*, 8 (3), 110-121.

Judson, E. (2006). How teachers integrate technology and their beliefs about learning: Is there a connection? *Journal of Technology and Teacher Education, 14*(3), 581-597.

Kagan, D. M. (1992). Implications of research on teacher belief. *Educational Psychologist*, 27(1), 65–90.

Lakoff, G. (1993). The contemporary theory of metaphor. In A. Ortony (Eds.), *Metaphor and thought* (pp. 202–251). Cambridge: University Press.

Lakoff, G., & Núñez, R. E. (2000). *Where mathematics comes from: How the embodied mind brings mathematics into being*. New York: Basic Books.

Lerman, S. (2002). Situating research on mathematics teachers' beliefs and on change. In G. C. Leder, E. Pehkonen & G. Törner (Eds.), *Beliefs: A hidden variable in mathematics education* (pp. 233–243). Dordrecht: Kluwer.

Levin, T., & Wadmany, R. (2006). Teachers' beliefs and practices in technology-based classrooms: A developmental view. *Journal of Research on Technology in Education*, *39*(2), 157.

Mishra, P. & Koehler, M. J. (2006). Technological pedagogical content knowledge: A framework for teacher knowledge. *Teachers College Record*, 108(6), 1017- 1054.

Noyes, A. (2006). Using metaphor in mathematics teacher preparation. *Teaching and Teacher Education*, 22, 898–909.

Pierson, M. E. (2001). Technology integration practice as a function of pedagogical expertise. *Journal of Research on Computing in Education*, 33(4), 413-429.

Reeder, S., Utley, J. & Cassel, D. (2009). Using metaphors as a tool for examining preservice elementary teachers' beliefs about mathematics teaching and learning. *School science and mathematics*, 109(3), 290–297.

Shulman, L. S. (1987). Knowledge and teaching: Foundations of the new reform. *Harvard Educational Review*, 57, 1-21.

Totter, A., Grote, G., & Stütz, D. (2006). ICT and schools: Identification of factors influencing the use of new media in vocational training schools. In *Proceedings of the 5th European Conference on e-Learning* (p. 469). Academic Conferences Limited.

Veen, W., (1993). The role of beliefs in the use of information technology: implications for teacher education, or teaching the right thing at the right time. *Journal of Information Technology for Teacher Education*, *2*(2), 139–153.

White, A. L. (2004). Can graphics calculators change pedagogical practices in secondary mathematics classrooms. In *Proceedings of 9th Asian Technology Conference in Mathematics* (pp. 153-160).

Wilson, M. S., & Cooney, T. J. (2002). Mathematics teacher change and development. In G. C. Leder, E. Pehkonen & G. Torner (Eds.), *Beliefs: A hidden variable in mathematics education* (pp. 127–147). Dordrecht: Kluwer. http://books.google.fi/

Yocum, K. (1996). Teacher-centered staff development for integrating technology into classrooms. *Technological Horizons in Education*, 24(4), 88–91.

A Study of Mathematics Teachers Conceptions of Their Own Knowledge of Technological Pedagogical Content Knowledge (TPACK)

Maria Sundberg

Dalarna University

Content

1 Introduction 160
2 Background 161
3 Method 162
4 Result 164
5 Discussion 166
6 References 167

Abstract

In this study, a sample of mathematics teachers at upper secondary level rated their knowledge with respect to the key domains described by the Technological Pedagogical Content Knowledge (TPACK) framework. The results indicated that teachers expressed that they had a high level of knowledge in terms of pedagogy and content and the combination of these, but the knowledge level was lower in terms of technology such as software installation or troubleshooting of computers. The results also indicated that there were small differences in the expressed level of knowledge between sexes and years of teaching experience. The study indicated that effective integration of digital tools should include training both in the educational use and the actual operation of the tools.

1 Introduction

In society today, we use a lot of digital equipment, such as computers, iPads and smart phones. This equipment is used in various forms such as forums, searching for facts, keeping in contact, etc. According to Säljö (2010), digital technology, has in recent decades seen enormous development. Computers have gained immense storage capacity through faster processors and well-developed software. The trends in education also show that the use of Internet and the availability e.g. through smart phones will be increasingly demanded (The Horizon Report, 2011). The changes society has undergone, on the digital side, the last few years' means that teachers are facing all kinds of digital tools related to their work. While it is not certain that computers and digital technology alone can improve teaching; good teachers also need to be available (Säljö, 2010). Todays schools compete with a variety of information and knowledge channels such as TV, video and computer games, the Internet, apps, etc. It could be argued that teachers should at least know about the different kinds of learning opportunities that are available for students in a digital world.

In Sweden, the curriculum for mathematics at upper secondary level says:

> Teaching should also give students the opportunity to develop their ability to use digital technology, digital media, and other tools which can occur in subjects typical of programmes. (Swedish National Agency for education, 2012, p. 1).

This means that digital tools should be a part of mathematics education in Sweden. NCTM also recognizes ICT as a part of education:

> Technology is an essential tool for learning mathematics in the 21st century, and all schools must ensure that all their students have access to technology (NCTM, 2008, p. 1).

As previously stated, access to technology is not enough. How the technology is used is dependent on the teachers and their knowledge. This has been pointed out by Drijvers:

> ...the teacher has to orchestrate learning, for example by synthesizing the results of technology-rich activities, highlighting fruitful tool techniques, and relating the experiences within the technological environment to paper-and-pen skills or to other mathematical activities (Drijvers, 2012, p. 12).

This means that teachers should integrate knowledge of student thinking and learning strategies along with knowledge of the subject with the use of digital tools in their teaching. This is a relatively new area of research in Sweden and the question is whether today's teachers have the skills to integrate knowledge of student thinking and learning strategies, knowledge of the subject and with digital tools.

The study is a replica of the study that Archambault and Crippen (2009) did in the USA which examined a national sample of 596 K-12 online teachers and measured their knowledge with respect to three key domains as described by the TPACK framework. Here, I will specifically examine how teachers assess their own competence in integrating pedagogy, mathematics and digital tools in their teaching practice. The research questions were,

- What is the perceived level of knowledge among teachers in mathematics that is specific to digital tools, pedagogy and subject content, including combinations of these?

- What differences are there in the perceived level of knowledge among teachers in mathematics that is specific to digital tools, pedagogy and subject content, including combinations of these, depending on the factors gender and teaching experience?

2 Background

The concept of pedagogical content knowledge (PCK) was introduced by Shulman (1986). He raised the issue of the need for a more coherent theoretical framework with regard to what teachers should know and be able to do. Mishra and Koehler (2006) built on Shulman's notion of PCK to articulate the concept of technological pedagogical content knowledge here referred to as TPACK.

The framework TPACK consists of three areas of knowledge: content knowledge (the topic to be taught), pedagogical knowledge (process and / or methods of learning and teaching), and technological knowledge (both ordinary such as the blackboard and more advanced such as computers) including the connections between these areas, (Mishra & Koehler, 2006) as illustrated in Figure 1 below. The relationship between these fields of knowledge is complex and nuanced (Mishra & Koehler, 2006; Schmidt et.al, 2009). For further definitions of each area of TPACK, see Koehler & Mishra (2008).

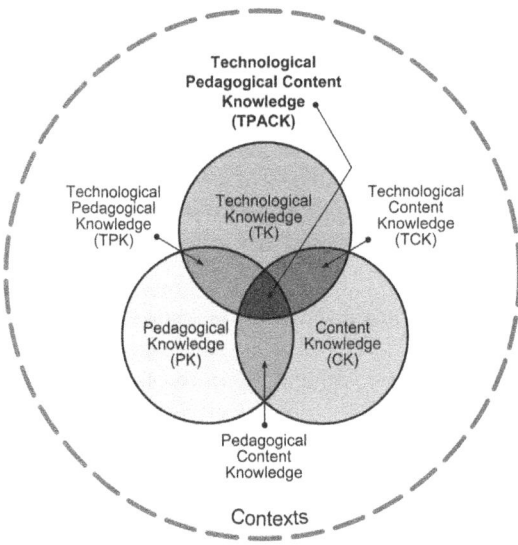

Figure 1 Framework of TPACK (Koehler, u.d.)

Archambault & Crippen (2009) studied TPACK in K-12, online teachers. Their results indicated that teachers perceived themselves to be proficient in the areas of pedagogy, subject content and the combination of these. However when it came to technology, the teachers expressed that they had less knowledge. The study also showed strong correlations between pedagogy and subject content, whilst the relationship between technology and pedagogy, and technology and subject matter was weak. The responses to the open-ended questions revealed difficulties with learning the new technology (Archambault & Crippen, 2009). Similar views have been observed in Turkish prospective primary teachers regarding the use of computers in mathematics education (Doğan, 2012). Although the Turkish teachers stated that the use of computers can help them to teach mathematics, they did not feel confident about it.

3 Method

The data collection was done through a web survey where the responses were automatically anonymous. That meant that the questionnaire was easy to administrate and compile. The disadvantages of online surveys is the same as traditional surveys, i.e. sources of error due to non-response, sampling, coverage and measurement errors are the same (Avery, 2006).

The questionnaire was a translated and modified version of that used by Archambault & Crippen (2009). Modifications were made to satisfy Swedish conditions and the level of the school system to be investigated (upper secondary school). A number of background questions were added, such as gender, university education and teaching experience. The questions were initially constructed using the areas of TPACK and the 23 questions were grouped in these areas (see table 1). The opportunity to comment on each section was given in open questions. The responses were given in a 5-point Likert scale, ranging from 1 = very poor to 5 = very good. A pilot study was made to test the instrument.

Table 1 The structure an variation in the questionnaire.

Area	Part	Question	Variation
Pedagogical knowledge (PK)	A	1-3	0.6
Technological knowledge (TK)	B	4-6	1.04
Content knowledge (CK)	C	7-9	0.6
Pedagogical Content knowledge (PCK)	E	13-16	1
Technological Content knowledge (TCK)	D	10-12	1.08
Technological Pedagogical knowledge (TPK)	F	17-20	1.28
Technological Pedagogical Content knowledge (TPACK)	G	21-23	1.16

These are some sample statements from the survey:

    A. My ability to adapt teaching method to the students' knowledge

    B. My ability to help students troubleshoot technical problems on their computers

    C. My ability to decide what mathematical concepts and how they should be taught in my class

    D. My ability to use different software in teaching (eg, Word, Powerpoint, Excel, etc.)

    E. My ability to support students in noticing connections between different concepts

    F. My ability to use different approaches to teaching through digital tools

    G. My ability to teach so that students achieve proficiency in digital software module specified in the curriculum and syllabus

The schools participating in this study were located in a medium sized state located in central Sweden. The selected schools where chosen based on their presentation of themselves where the criteria was stressing the use of ICT in education. 13 schools were chosen from an initial group of 25. In these 13 schools there were 71 mathematics teachers. Due to technical problems, 7 teachers couldn't participate in this study leaving the maximum number of potential respondents to 64. Of these, 26 replied giving a response rate of 41 %. Mean value and standard deviation was calculated for each question. Due to the small number of respondents, no further statistical analysis was made and the results were compared with previous research. This study is descriptive and the result cannot be generalized beyond the response group. However general tendencies can be indicated.

4 Result

The results are summarized and presented by the different areas of TPACK and from now on I will use the abbreviations for the different areas. First we look at the differences between the areas of PK, CK and PCK that are higher than the other TK, TCK, TPK and TPACK, see Table 2:

Table:2 Descriptive results within the areas concerning teachers' estimadted ability in different situations.

Area	Mean value	Standard deviation
Pedagogical knowledge (PK)	3.960	0.701
Technological knowledge (TK)	3.160	1.233
Content knowledge (CK)	4.227	0.741
Pedagogical Content knowledge (PCK)	4.250	0.766
Technological Content knowledge (TCK)	3.747	1.034
Technological Pedagogical knowledge (TPK)	3.360	1.063
Technological Pedagogical Content knowledge (TPACK)	3.347	1.149

Table 2, also shows that the spread among individual teachers is greater in the areas of TK, TCK, TPK, and TPACK than in the areas of CK, PK and PCK.

In all areas except CK and PCK (which is basically the same), men rated their knowledge higher than women do, although the differences are quite small (Table 3). The largest gender difference exists within the subarea TK where the values differ by more than 1 unit, including TCK (0.7 units), TPK (0.6 units) and TPACK (0.3 units).

Table 3 Descriptive results for issues concerning teachers' estmated ability in different situations by gender

Area	Mean value	
	Women	Men
Pedagogical knowledge (PK)	3.923	4.000
Technological knowledge (TK)	2.564	3.806
Content knowledge (CK)	4.256	4.194
Pedagogical Content knowledge (PCK)	4.269	4.229
Technological Content knowledge (TCK)	3.410	4.111
Technological Pedagogical knowledge (TPK)	3.077	3.667
Technological Pedagogical Content knowledge (TPACK)	3.205	3.500

The groups of teacher experience were chosen based on the few respondents; smaller groups would have meant groups with no respondents in it. All areas show a maximum value for those with 10-20 years of teaching experience (table 4). The difference from 0-10 years to 10-20 years is greatest for the area TPACK (1.1 units) for other areas, the differences are between 0.3-0.8 units. The reduction from 10-20 years to 20 years-, is greatest for TPACK (1 unit).

However, as can be seen in table 4, the variations are quite small between the groups.

Table 4 Descriptive results for issues concerning teachers' estimated ability in different situations, sorted by teaching experience

Area	All	0-10 years	10-20 years	20- years
Pedagogical knowledge (PK)	3.960	3.718	4.333	4.111
Technological knowledge (TK)	3.160	2.949	3.778	3.000
Content knowledge (CK)	4.227	3.897	4.667	4.500
Pedagogical Content knowledge (PCK)	4.250	4.115	4.417	4.375
Technological Content knowledge (TCK)	3.747	3.564	4.111	3.778
Technological Pedagogical knowledge (TPK)	3.360	3.192	3.875	3.208
Technological Pedagogical Content knowledge (TPACK)	3.347	3.051	4.167	3.167

5 Discussion

This study has used TPACK as a framework for measuring the perceived level of knowledge of a group of mathematics teachers working in schools explicitly focusing on ICT and other digital tools, these teachers should have knowledge related to these areas. It turned out to be difficult and complex. The pilot study revealed difficulties in distinguishing the areas implying that TPACK as a framework has difficulties in measuring the level of knowledge in the various fields, but it can still be a framework that describes what a teacher needs to have knowledge about, what distinguishes teachers from educators or technicians. From this present study a few conclusions can be drawn.

In the study of what the perceived level of knowledge of mathematics teachers at secondary level is within the framework TPACK, teachers express that they have good or very good knowledge about traditional teaching (teaching that doesn't integrate digital tools). However, they are more insecure in their knowledge regarding technology and the integration of technology in teaching. There are also some differences among individual teachers. In the respondents' written comments there are two branches, those who want to try and integrate the technology but do not have time or feel they don't know enough about the technology and need to learn it first, and those who do not wish or need to use technology at all. As one of the respondents writes "You can view different representations of a concept even without the use of digital tools, it's called a review." This can be interpreted as an unwillingness to change that is one of the criteria for a mathematics teacher with good TPACK skills, according to Grandgenett (2008). The results of this study may be due to several factors, including what experience teachers have acquired in their professional activities. The survey shows a trend toward higher levels of knowledge in all areas after 10-20 years of teaching experience, which is supported by Samuelsson and Samuelsson (2011) and Nilsson (2010). Teacher education may also be a factor, if not previously learned to teach through digital tools, it can be hard to do it on your own unless the persons own interest is involved. According to Graham (2009), it is natural that the subareas with technology is lower than most but with practical training, these skills can increase. It could also be that the development in technology progresses so very quickly that it is difficult to "keep up". If an education today would contain "the latest" in technology and education, it would still be "old" when the student graduates. As Dewey (1916) says, "If we teach today as we taught yesterday, then we rob our children of tomorrow." By changing teacher training and providing appropriate technical experience can we improve mathematics education (Landry, 2010).

The results of this study are in line with the observations of Archambault & Crippen (2009), they are almost identical despite different contexts and respondents. It seems like it exist a general view about yourself and your perceived knowledge regarding the areas of TPACK.

As a teacher educator, we cannot assume that the pedagogical knowledge of ICT follow for instance the use of ICT or the other way around. We need to be more specific about how to use it, when to use it, and be able to say why we should use it.

6 References

Archambault, L. & Crippen, K. (2009). Examining TPACK among K-12 online distance educators in the United States. *Contemporary issues in Technology and teacher education*, ss. 71-88.

Avery, R. B. (2006). Electronic course evaluations: Does an online delivery system influence student evaluations? *Journal of Economic Education*, 21-37.

Dewey, J. (1916). *Democracy and education.* New York: Free Press.

Doğan, M. (2012). Prospective Turkish primary teachers views about the use of computers in mathematics education. Springer Science+Business Media B.V.

Drijvers, P. (2012). Digital technology in mathematical education: Why it works (or doesn't). *Proceedings of the 12th International Congress on Mathematical Education*. Seoul, Korea.

Graham, C. B. (2009). Measuring the TPACK confidence of inservice science teachers. *TechTrends*, 70-79.

Grandgenett, N. F. (2008). Perhaps a matter of imagination: TPACK in mathematics education. *Handbook of technological pedagogical content knowledge (TPACK) for educators* (ss. 145-166). New York: American Association of colleges for teacher education.

Koehler, M. J. (u.d.). *TPACK – Technological Pedagogical and Content Knowledge*. Received from http://tpack.org at 2011 28/4

Koehler, M. & Mishra, P. (2008). What is technological pedagogical content knowledge (TPCK)?. *Handbook of technological pedagogical content knowledge (TPCK) for educators* (ss. 1-58). New York: American Association of Colleges for Teacher Education.

Landry, G. A. (2010). *Creating and Validating an Instrument to Measure Middle School Mathematics Teathers' Technological Pedagogical Content Knowledge (TPACK)*. Tennessee: University of Tennessee.

Mishra, P. & Koehler, M. J. (2006). Technological Pedagogical Content Knowledge: A Framework for Teacher Knowledge. *Teachers College Record*, 1017-1054.

National Council of Teachers of Mathematics. (2008). The role of technology in the teaching and learning of mathematics. Retrieved from http://www.nctm.org/uploadedFiles/About_NCTM/Position_Statements/Technology%20final.pdf at 2013 22/4

Nilsson, S. (2010). Professional experience in industry an school: A study within learning situations an the industry programme at upper secondary school [Yrkeserfarenhet inom industri och skola: En studie av lärandesituationer på gymnasieskolans industriprogram]. Skövde: University of Skövde.

Samuelsson, M. & Samuelsson, J. (2011). The meaning of Professional experience: Self-assessment of teaching effectiveness of the Swedish teachers in textile crafts [Yrkeserfarenhetens betydelse: Självskattning av undervisningseffektivitet hos svenska lärare i textilslöjd]. *Didaktisk Tidskrift*, ss. 249-272.

Shulman, L. (1986). Paradigms and research programs in the study of teaching: A contemorary perspective. i M. C. Red. Wittrock, *Handbook of research on teaching* (ss. 3-36). New York: MacMillan.

Schmidt, D. A., et al. (2009). Technologycal Pedagogical Content Knowledge (TPACK): The development and validation of an assessment Instrument for Preservice teachers. *Journal of research on Technology in Education*, 123-149.

Swedish National Agency for education, (2012) Retrieved from http://www.skolverket.se/polopoly_fs/1.174554!/Menu/article/attachment/Mathematics.pdf at 2013 22/4

Säljö, R. (2010). Digital tools and challenges to institutional traditions of learning: technologies, social memory and the performative nature of learning. *Journal of computer assisted learning.*

The Horizon report. (2011). Retrieved from http://wp.nmc.org/horizon2010/ at 2011

Why Do Experts Pose Problems for Mathematics Competitions?

Igor Kontorovich

Technion – Israel Institute of Technology

Content

1 Introduction and Theoretical Background 172
2 Method 173
3 Findings 174
    3.1   Experts' Intellectual Need 174
    3.2   Experts' Socio-Psychological Needs 176
4 Discussion 179
6 References 181

Abstract

Inspired by the recurrent findings on the steady decrease in students' interest in mathematics, this paper is concerned with sources of experts' motivation for posing problems for mathematics competitions. Twenty-six experts from nine countries participated in the study. The inductive analysis of the data suggests that experts utilise posing problems for mathematics competitions for fulfilment of their internal needs: an intellectual need for enriching their mathematical knowledge base and a socio-psychological need for belonging, recognition and appreciation. Educational implications of the findings for students are discussed and future research directions are presented.

1 Introduction and Theoretical Background

Motivation has been acknowledged as one of the key factors in teaching and learning. Empirical research conducted in mathematics education has shown that motivation is closely related to students' performance, achievements and beliefs (e.g., Lewis, 2013; Middleton & Spanias, 1999). Moreover, motivation is a part of the students' scholastic mathematical experience. This experience plays a significant role in a student's decision to learn mathematics courses in the future and pursue a career in mathematics (Middleton & Spanias, 1999). In light of the above, the recurrent findings on the steady decrease in students' motivation and interest in mathematics are truly disturbing (e.g., Lewis, 2013; Middleton & Spanias, 1999).

In order to develop new insights into sources of students' motivation, this paper focuses on experts in mathematics. This focus is in line with the established practice of using research on experts as a source of ideas for fostering mathematics learning for novices (see a detailed discussion on this point in Kontorovich and Koichu, in press). Several categories of experts were considered for the purposes of this study: practicing mathematicians, lecturers, teachers, task designers and mathematics textbook writers. Eventually, the decision was made to recruit expert problem posers (EPPs) for mathematics competitions (MCs). The decision was based on the assumption that in many cases EPPs for MCs receive neither fiscal rewards nor academic accreditation for being engaged in mathematics, unlike the aforementioned categories of experts. Thus, identifying the sources of the EPPs' motivation could shed new light on intrinsic motivation in engaging in mathematics.

The preliminary study with 22 adult participants from the Competition Movement (i.e. coaches, EPPs and organizers of MCs) was presented at the 17th MAVI conference (Kontorovich, 2012). The findings of the preliminary study suggested that the participants shared a pedagogical agenda consisting of four interrelated goals: to provide students with opportunities to learn meaningful mathematics, to strengthen their positive attitudes towards mathematics, to create cognitive challenges for the students and to surprise them. Competition problems were perceived by the participants of the preliminary study as a means of achieving these goals.

In light of the above, it can be suggested that fulfilling this pedagogical agenda is a possible source of experts' motivation to pose competition problems. However, EPPs might also consider alternative ways of engaging in students' mathematical education. For instance, they could publish books or teach mathematics at school. The experts' decision to invest their time and effort in creating problems for MCs implies that they may possess additional motivational sources. Identification of these sources is the goal of the current study.

Lewis (2013) wrote that research on motivation in mathematics education is frequently concerned with attitudes and that it is undertaken from a statistical point of view. He reminds us that motivation is an emotional phenomenon, and therefore phenomenological methods propose an added value in its exploration. Lewis's approach has been adapted in the current study. Namely, the study is conducted using qualitative methods, and motivation is analysed from the participants' point of view, i.e. people who actually possess it.

2 Method

Twenty-five men and one woman participated in the study. All the participants are active problem posers for national, regional and international MCs, such as The Baltic Way, Mathematical Kangaroo and Tournament of the Towns. Twenty-three experts were above the age of 40, and the average age of the experts was 50 (SD=12). The experts' problem-posing experience ranged from seven years to more than 30. The academic background of the experts is as follows: one expert holds a master's degree in mathematics education; six received a master's degree in mathematics; two have a PhD in mathematics education; and 17 experts hold a PhD in mathematics.

Nineteen experts are employed in colleges and universities; one expert teaches in a high-school and another works in the high-tech industry. One expert has an official paid position in the organizational committee of national competitions. Only one expert (Olivia) makes her living posing riddles for daily newspapers and publishing riddle books. Thus, the study's a-priory assumption that problem posing for most of the experts is a non-profitable practice proved to be a correct one.

The participants reside in Australia, Bulgaria, Israel, Latvia, Lithuania, the Russian Federation, Spain, Sweden and the USA. Eighteen experts chose to interact with me in Russian, 4 preferred English and the remaining 4 preferred Hebrew. Thus, the majority of the experts participating in this study were influenced, or nurtured, by the [former] Soviet mathematics education. In other words, the variety of the experts' countries of residence does not fully represent the diversity of mathematical views and perceptions.

The data were collected using semi-structured in-depth interviews (e.g., Evans, Patterson & O'Malley, 2001) and open questionnaires. The Israeli participants were interviewed face-to-face. Nine participants from other countries were interviewed by telephone or Skype. The interviews lasted between 60 and 90 minutes. The questions were sent to participants by e-mail a week before the interview. This was done in order to "prep" the participants for the forthcoming interview. All the interviews were audio-taped and transcribed. The remaining eight experts preferred to answer in writing by filling in a questionnaire consisting of the central questions of the interview. After a preliminary analysis of the data the experts were then asked to take part in follow-up interviews and to answer a set of question for clarification and validation. Nine experts consented to take part in these follow-up activities.

In both the interviews and the questionnaires, the experts were asked to tell about the role that the competition movement plays in their life. They were asked specifically about their problem-solving and problem-posing experience, their reasons for joining the competition movement in the first place, and their motivation for staying in the movement and becoming problem posers. The data were analysed using an inductive approach in order "[...] to allow research findings to emerge from the frequent, dominant or significant themes inherent in raw data, without the restraints imposed by structured methodologies" (Thomas, 2006, p. 238).

3 Findings

3.1 Experts' Intellectual Need

According to Wikipedia (n.d.) the *intellectual need* of a person is a special type of intrinsic motivation for gaining a new piece of knowledge. In the excerpts below, two of the study participants, Ari and Mike (pseudonyms), explicitly refer to their "built-in" curiosity and desire to learn new mathematics:

> Ari: All my life I was aware that the only things I need in my free-time are a pen and a sheet of paper. I am always preoccupied with some geometrical configurations and their properties... which is why I usually have too many questions about them yet very few answers. I think this is the fate of any researcher. [...] My motivation for conducting [mathematical] research is curiosity and the desire to reveal what is true: [For instance,] Is some property generalizable from a special case to other cases as well? Does a particular configuration conceal any additional properties that we are not familiar with, or have we reached the maximum? [Translated from Hebrew.].

> Mike: When I pose a problem, I learn [i.e. acquire new mathematical knowledge]. Actually, it is one of my character traits. I don't even understand where it comes from. Every time I see a problem that interests me, I feel a powerful desire to "strengthen" it, to take it one step forward, to learn new things about it. In this way many new problems are created.

> Interviewer: What motivates you "to strengthen the problems"?

> Mike: An ambition to know how this world works. Why does a problem include a particular given, rather than another one? The discovery and understanding create an incredible feeling, which gives you wings! [Translated from Russian.].

Two phenomena can be observed in Ari's and Mike's statements. First, they both describe their desire to learn new mathematics in a highly emotional manner and report that its fulfillment creates a pleasant and rewarding feeling. Second, Ari and Mike refer to this desire as a significant characteristic of their personality rather than a situational feeling. The intensiveness and ubiquity of this desire bear resemblance to the physiological and socio-psychological needs described by Maslow (1943), which facilitates referring to this burning desire as an *intellectual need for enriching their mathematical knowledge base*.

Ari and Mike exposed their intellectual need more elaborately than the rest of the study participants. In the data collected on 10 additional experts, intellectual need was noted in certain sentences and phrases, in which the experts referred to their aspiration to acquire new mathematical knowledge. For instance: "From the moment I remember myself, I always wanted to understand how…", "I was always curious about …", "There are problems and questions that will keep you up all night!". Note that in the first two phrases, the idiosyncrasy of the need is emphasized (like in the case of Ari and Mike), when in the third phrase the need is described as contextual, i.e. being provoked by a particular mathematical situation. This type of situation is exemplified below in the case of Alexis and Mavlo's Theorem:

Mavlo's Theorem states that one of the arcs created by a side of a triangle in the Euler circle equals the sum of the arcs created by the other two sides. In some cases, the arcs are created by extending the sides of the triangle up to the intersection with the Euler circle. When Alexis was asked to recall his first reaction to the theorem, he said:

> "My intuition told me that it is possible to take this theorem one step forward. [...] The Euler circle is the most familiar one in geometry and it leaves other circles a long way behind. I wanted to find out if some other circles also have a chance to stand out."

After that, Alexis was engaged in a search for other circles with properties related to arcs created by triangles. During his search, he acquired mathematical concepts and theorems which were new to him, and he posed problems based on his new knowledge. One of these problems appeared in 2010 in the Russian Geometry Olympiad named after Sharigin. In addition, Alexis described his mathematical journey of knowledge acquisition in a self-reflective paper (see Miakishev, 2010).

Similarly to Alexis, other experts also said that when they pose problems for competitions, they frequently get exposed to mathematics which is new to them, generate questions that they did not think about before and encounter problems they are unfamiliar with. Therefore, it can be deduced that problem posing for MCs is fertile ground for fulfilling the experts' intellectual need for new mathematical knowledge.

It is noteworthy that EPPs are not interested in just any piece of mathematical knowledge which is new and unfamiliar to them. The analysis showed that each participant possesses 1-3 fields of mathematical interests and seeks to enrich his or her knowledge base in these fields. Examples of such fields are Euclidean geometry, 3-D geometry, number theory, logic problems, linear algebra etc.

3.2 Experts' Socio-Psychological Needs

The experts tend not only to discover new [for them] mathematical facts, but turn these facts into problems and propose them to various MCs. This behavior cannot be attributed to the aforementioned intellectual need. From the perspective of experts' pedagogical agendas, it could be said that experts use these problems to achieve their goals regarding the students who participate in the competitions (see Introduction and Theoretical Background). In this sub-section I argue that the act of publishing the created problems in MCs can also be interpreted as a means for fulfilling experts' *socio-psychological needs for belonging, recognition and appreciation*. To recall, this need is included in Maslow's (1943) pyramid and its fulfillment is acknowledged by many schools as a powerful motivating factor (e.g., Maccoby, 1988; McClelland, 1971).

The importance of the sense of belonging in the case of EPPs is implied by the fact that all the experts who participated in the study were involved in MCs when they were students. Consequently, the experts have belonged to the competition movement for the greater part of their lives, when at some point they made "a switch" from solvers to posers. When asked why they decided to stay in the competition movement as adults, some of the experts explained their decision using the law of inertia: "I got used to Olympiads; it was my hobby", "I like dealing with mathematics, so how else am I supposed to do it?!". One of the experts, Ronny (pseudonym), gave a more detailed explanation:

> "It was when I started studying in college. It was a very stressful time for me: I finished high school, moved out of my parents' house and relocated to another city; I was away from my family. The studying [in college] was so different from what I was used to in high school. The mathematics was so different... So I looked for something to lean on, something that was familiar and natural to me." [Translated from Hebrew.].

Another expert, Olivia, explained her decision to remain in the competition movement metaphorically:

> "I participated in competitions when I was a student, but I never excelled in them. And I always had a strong desire to be noticed. This desire gave me the motivation to pose problems. I think that the desire to be better than others in something is natural. For instance, if someone likes ice skating but he isn't a good ice skater, he can

become a good coach, or a judge or an ice show producer." [Translated from Russian].

In this quote, Olivia admits that she possesses the need for recognition and appreciation, and this explains her decision to remain in the competition movement. Namely, Olivia puts forward the connection between her modest achievements in solving competition problems and her decision to become a problem poser: the fact that she was a part of the competition movement in the past and that she is still a part of it today enables her to compensate for the lack of recognition and appreciation back then with the recognition and appreciation she gets today, thanks to the high-quality problems she creates.

Olivia's idea about the connection between posing problems and gaining recognition and appreciation can also be extracted from Leo's words:

> "It is important that Olympiad questions include as many ideas as possible. If all the questions are similar to each other or to the questions that appeared in the previous Olympiads, it can create an impression that we [problem posers] don't have enough ideas". [Translated from Russian.].

It can be concluded from Leo's words that he perceives the competition problems as a two-sided assessment tool: It enables one to assess the problem-solving skills of the students as well as the problem-posing skills of the experts. Therefore, when posing a problem, Leo makes an effort to create a positive impression with his problems on the intended solvers, fellow problem posers, coaches, organizers of the competitions etc.

In 25 (out of 26) cases, the needs for belonging, recognition and appreciation went hand in hand. The excerpt from the case of Mike presented below shows that these emotions can intertwine in a complex manner. Mike specializes in the field of algebraic inequalities. For more than 20 years he has been solving these types of inequalities, collecting inequalities that appeared in MCs over the world, posing his own inequality problems and even developing methods for solving and proving inequalities.

Mike said that not many people are as interested in inequalities as he is. Moreover, over the years he has reached extraordinary levels of expertise in this field and asserted that he has difficulty finding people who are able to understand and appreciate his findings. He found these people on the web-forum "Art of Problem Solving" and became an active member with more than 10,000 posts. Thus, it can be concluded that the forum serves for Mike as a platform for having meaningful interactions with like-minded others and it provides him with a sense of belonging.

However, it turned out that Mike conducts his mathematical research in writing and it is summarized in twenty-two thick notebooks. The following excerpt from the interview with Mike focused on this phenomenon:

Mike: If all my stuff was saved in the computer, someone would steal it for sure. I don't want this to happen: it is an enormous amount of work, the work of all my life actually! Why should I give it to someone else?!

Interviewer: Why you are so sure that someone would "steal" your work?

Mike: It's not paranoia, it has already occurred in the past.

Interviewer: Can you tell me about this?

Mike: Naughty XXXians [names a country] stole one of my problems [he says the words with an intonation of pride, smiles and giggles at this point]. I posted one of my problems on the forum "Art of Problem Solving" and a year and a half later, it appeared in the XXXian Olympiad.

Interviewer: And how did it make you feel then?

Mike: Well, you know... It is important to me that people see my results. If someone steals them, well... So be it! If I want, I can always prove that I was the first one to publish them.

Mike's case exemplifies the importance of the sense of belonging, recognition and appreciation in expert problem posing. Similarly to Olivia and Leo, Mike wants people to attribute their positive emotions provoked by mathematical problems to him, the person who created these problems. This idea can be extracted from Mike's usage of the word "naughty": While, on the one hand, he claims that some people stole his problem, on the other hand, they popularized it and, therefore, automatically popularized its creator. Interestingly, it is enough for Mike to know that he is the creator of the popularized problem, even if people are not aware of this fact. Indeed, he did not try to expose himself as the real creator of the problem; he is satisfied simply knowing that the exposure is achievable. Regarding the act of "stealing" his intellectual property (from his point of view), apparently, this is the "price" he is willing to pay for the sense of belonging which he is granted by the forum.

As can be seen from the excerpts above, experts' socio-psychological needs were observed in very personal and revealing materials, which they chose to share. Apparently, this special kind of material could not appear in the data on all the participants of the study. This fact partly explains why evidences of experts' socio-psychological needs were observed only in five cases. In addition, it turned out that in four of these cases, experts regularly publish their mathematical results in special journals and handbooks. This action can be interpreted as a realization of a high degree of need for recognition and appreciation, a degree which could not be completely fulfilled by publishing problems in competitions. The data collected on eight additional experts included phrases such as "I like sharing new mathematical results with people" or "I'm very proud, when people say 'wow!' about my problems". These phrases can also be seen as evidence of experts' needs for recognition and appreciation.

4 Discussion

The main contribution of this paper is in its empirical identification of two motivational sources among experts in mathematics: an intellectual need for enriching their mathematical knowledge base and a socio-psychological need for belonging, recognition and appreciation. The identified sources serve as an explanatory framework for understanding the practices of 26 Expert Problem Posers (EPPs) who create problems for Mathematics Competitions (MCs).

The analysed case of EPPs and MCs is an example of a "good match" between individuals who possess particular needs and a framework for participation that addresses these needs and enables its participants to create products of a high quality, develop their knowledge and skills, and actualize their capabilities and talents. Therefore, creating similar frameworks for learning purposes can be considered one of the central aims of education. Moreover, I argue that intellectual and socio-psychological needs are typical to all people, and therefore, they could be and should be utilized for designing these frameworks.

The educational community has been extensively utilizing students' intellectual needs for learning mathematics. For instance, Harel and his colleagues introduced and empirically based a comprehensive framework for the instruction of mathematics (e.g., Harel, 2008). According to it, learning of a new construct of knowledge occurs as a result of students being engaged in carefully designed tasks, the solution of which requires the missing knowledge. However, Harel's approach to the concept is purely epistemological. In his words:

> "There is often confusion between *intellectual need* and *motivation*. The two are related but are fundamentally different. While intellectual need belongs to epistemology, motivation belongs to psychology. Intellectual need has to do with disciplinary knowledge being born out of people's current knowledge through engagement in problematic situations conceived as such by them. Motivation, on the other hand, has to do with people's desire, volition, interest, self determination, and the like." (Harel, 2008, p. 897-898).

The data presented in the current paper show that in the case of EPPs, intellectual need is one of the most powerful motivational sources deriving from their personalities. For them it is a state of mind that enables them to sense the need for a nonspecific piece of mathematical knowledge without being involved in problematic situations. This finding is in line with Ericsson (2006), who argued that experts tend to engage in deliberate practices (problem posing in this case) in order to extend their already well-developed knowledge bases and to sharpen their professional skills.

In light of the above, there is room for rigorous empirical investigation of students' personal intellectual needs, preferences and interests, for structuring them in clusters and for designing learning settings that will match the identified clusters. Note that nowadays, the idea of a learning setting which is matched to the intellectual cluster of a student is realized in terms of a course in some mathematical area with a particular level of complexity (e.g., Algebra 3). In other cases the course is merely adjusted to the age of the students (e.g., Mathematics for the 9th grade). The aim of the proposed investigations is to address the intellectual needs of particular students with narrower aspects of the learning setting, such as mathematical topics that will interest them, problems that will challenge and engage them, exercises that will promote competencies without boring them etc. Matching students to a carefully designed learning setting can be a powerful tool aimed at unlocking their mathematical potential.

Regarding the socio-psychological need, recognition and appreciation are typically achieved through high grades, which are assigned by figures of authority. The grades are transformed into a goal to achieve, while learning is solely the means to an end. The situation is significantly different in the case of EPPs: *First*, they are rewarded by being accepted into the elite society of problem posers for prestigious MCs. *Second*, rewarding is based on high-quality products (excellent mathematical problems in this case) produced by the experts. Since these products are inseparable from learning mathematics, a direct connection is established between learning and reward. *Third*, the experts are rewarded by colleagues and consumers of their products – principally the students who participate in MCs. In this way, the rewarding communities encourage the posing of high-quality problems, and thus they enhance learning.

Research has been relatively silent about empirical attempts to model these communities in a classroom setting (see Brawn & Walter, 1996 for a rare example). Therefore, there is room for additional inquiry of students' socio-psychological needs and for the ways these needs can be addressed when learning mathematics. I believe that rigorous study of the movement of Mathematics Competitions is an insightful step for addressing these issues.

5 Acknowledgements

I wish to extend my sincere thanks to the participants of the study and to my academic advisor Professor Boris Koichu for his insightful comments and support.

6 References

Brown, S. I., & Walter, M. I. (1996). In the classroom: Student as author and critic. In Brown, S.I. & Walter, M.I. (Eds) *Problem Posing: Reflections and Applications* (pp. 7-27). Hillsdale, NJ: LEA.

Ericsson, K. A. (2006). The influence of experience and deliberate practice on the development of superior expert performance. In K. A. Ericsson, N. Charness, P. Feltovich, & R. R. Hoffman (Eds.), *Cambridge Handbook of Expertise and Expert performance* (pp. 685-706). Cambridge, UK: Cambridge University Press.

Evans, M., Patterson, M., & O'Malley, L. (2001). The direct marketing- direct consumer gap: qualitative insights. *An International Journal*, 4(1), 17-24.

Harel, G. (2008). A *DNR* perspective on mathematics curriculum and instruction. Part II: with reference to teacher's knowledge base. *ZDM*, 40, 893-907.

Kontorovich, I. & Koichu, B. (2012). Feeling of innovation in expert problem posing. *NOMAD*, 17(3-4), 199-212.

Kontorovich, I. (2012). What makes an interesting mathematical problem? A perception analysis of 22 adult participants of the competition movement. In B. Roesken & M. Casper (Eds.), *Proceedings of the 17th MAVI* (Mathematical Views) Conference (pp. 129-139). Bochum: Germany.

Lewis, G. (2013). A portrait of disaffection with school mathematics: The case of Anna. *Journal of Motivation, Emotion, and Personality*, 1, 36-43.

Maccoby, E. E. (1988). Gender as a social category. *Developmental Psychology*, 24(6), 755-765.

Maslow, A. (1943). A theory of human needs. *Psychological Review*, 50, 370-396.

McClelland, D. C. (1971). *Assessing Human Moti*vation. Morristown: General Learning Press.

Miakishev, A. (2010). Progulki po okrugnostiam: ot Eilera do Teilora [Circle trips: from Euler to Taylor]. *Matematika: Vsei dlia uchitelia*, 6(6), 1-16.

Middleton, J., & Spanias, P. (1999). Motivation for achievement in mathematics: Findings, generalizations, and criticisms of the research. *Journal of Research in Mathematics Education*, 30(1), 65-88.

Thomas, D. T. (2006). A general inductive approach for analyzing qualitative evaluation data. *American Journal of Evaluation*, 27(2), 237-246.

Beliefs and Mathematical Reasoning during Problem Solving across Educational Levels

Ioannis Papadopoulos
Department of Primary Education
Aristotle University of Thessaloniki, Greece

Content

1   Introduction 184
2   Some Theoretical Aspects 184
    2.1   Reasoning 184
    2.2   Beliefs 185
3   Description of the Study 185
4   Results and Discussion 187
5   Conclusions 193
6   References 194

Abstract

In this paper the status of empirical mathematical reasoning during problem solving across primary, secondary and tertiary education is studied. The main aim is to see whether the very same beliefs influence the students' performance in the same way across educational levels. The results show that despite sharing the same beliefs, the way these beliefs affect students' performance (positively or negatively) is different for different ages. More precisely, as we move from primary grades to college, the students' ability to employ empirical mathematical reasoning is inclined as they persist to ask for connections with more formal ways of working. Even though the students solved the same task and shared the same beliefs, the negative effects of these beliefs were stronger for older students.

1 Introduction

The logically correct reasoning was an achievement developed and tested through history. This constitutes a unique contribution of mathematics to the culture of science and we have to be very careful to preserve it. To obtain this, a culture of argumentation is needed in the mathematics classroom from the primary grades up all the way through college (Ball et.al., 2002). There is evidence (Maher & Martino, 1996) that young students can be capable of engaging in reasoning which varies from empirical arguments to abstract. However, this ability is influenced positively or negatively by the solver's beliefs. It seems that the way you think about a mathematical topic affects the way you approach it, e.g. the fact that the students view mathematics mostly as a set of rules and procedures to be learned by rote (Crawford et.al., 1994). Thus, accepting that beliefs might be misleading for the students no matter the age or grade, the question we try to answer is: Is there any differentiation across grades and/or educational levels? Is this influence - caused by beliefs - stronger for some ages than for others? This is why we study three populations (primary, secondary and tertiary education) asking for a comparison of the students' reasoning as it may be influenced by their beliefs, given that they cope with the same task.

2 Some Theoretical Aspects

2.1 Reasoning

Any type of mathematical reasoning suggests an a-posteriori summation of the lines of thought taken before in order to produce assertions and reach conclusions in task solving (Mamona-Downs & Downs, 2013). Reasoning is not necessarily based on formal logic and it may be incorrect (Lithner, 2008). In this respect Lithner makes a distinction between *reasoning* and *argumentation* and takes argumentation as the "substantiation" that convinces you that the "reasoning is appropriate". In the context of problem solving this means that the students rely on mental images and loosely grounded representations, they initialize, collate and monitor the argumentation involved. In other words, this refers to what children at a particular grade level could be expected to know that could be used in a proof; what kind of arguments they are capable of making; and what kinds of representations they can use. Lithner (2008) describes two types of reasoning: (1) Creative Reasoning characterized by novelty, plausibility and mathematical foundation; it utilises resources, heuristics, beliefs, control skills and supporting environment (Schoenfeld, 1985), and (2) Imitative Reasoning which can be seen either as Memorised Reasoning (recalling an answer and writing it without consideration) and Algorithmic Reasoning (recalling a certain algorithm and implementing the rules. According to Hanna (2000) the ability of young students to organize arguments into a logical chain grows in sophistication as the learner matures. The research literature presents some examples. Maher and Martino (1996) describe 4th graders presenting a version of proof by cases (or proof by exhaustion) and proof by mathematical induction during a combinatorial problem task. Ball (in Ball et al. 2002) describes how students in grade 3 use diagrams (visual aid) to 'prove' that 'the sum of two odds is even'. Finally, retelling the story of what has been done is another way of reorganizing the ideas into a chain of logically connected statements (Fosnot & Jacob, 2010). For students this means that they need not only to convince, but also to explain. English (1997) suggests that elementary school textbooks and curricula must take reasoning processes into account, something that possibly could lessen the difficulties that secondary school students face because of the abrupt introduction to the new requirement of proof.

2.2 Beliefs

Schoenfeld (1985) defines beliefs as "one's mathematical world of view, the perspectives with which one approaches mathematics and mathematical tasks. One's beliefs about mathematics can determine how one chooses to approach a problem, which techniques will be used or avoided, and so on" (ibid, p. 45). Sumpter (2013) uses Schoenfeld's definition as a starting point to conclude that beliefs essentially are the understanding that shapes the ways one conceptualizes and engages in mathematical behaviour and therefore in this sense beliefs are considered to be cognitive. Research findings show that students' beliefs can either facilitate or hamper their problem solving process. Schoenfeld (1992) claims that the consequences of the students' beliefs are extraordinary powerful and often negative. On the other hand Carlson (1999) investigating the mathematical behaviour of graduate students and assessing their beliefs found that these beliefs fostered positively their persistence and confidence. The results of these studies reflect the status of beliefs as far as a specific uniform population is concerned (i.e., the participants are of the same age or educational level). We try to broaden this scope by recording at the same time the influence of beliefs on different populations given that: (a) they represent different educational levels, and (b) they are coping with the same tasks.

3 Description of the Study

The participants were students from primary and secondary school as well as university students (prospective elementary teachers in the Department of Primary Education). More precisely there were 82 primary school students (37 grade 5 and 45 grade 6), 35 secondary ones (17 grade 10 and 18 grade 11) and 50 prospective teachers. There was no prerequisite mathematical knowledge for solving the task and this is why the participants' mathematical background is not described. However, it would be helpful to keep in mind that during their regular classes the secondary school students were oriented towards science and mathematics and also, we assume that the prospective elementary teachers do not have a similar mathematical background to other tertiary students (i.e., undergraduate mathematics). The task was comprised of two parts (Figure 1). For each part of the task, students had a separate session.

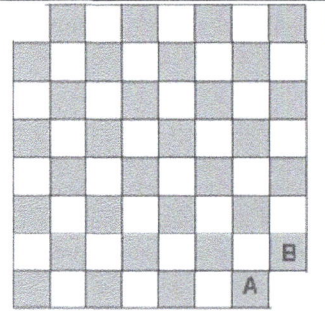

Part I. In this chessboard the bottom-right and the top-left squares are missing. Using dominos like the one given in the picture and applying the rules that govern domino try to find a path that starts from A and ends to B (or the opposite).

Part II. For the same chessboard and with the same white squares missing, try to cover the whole area using the given rectangular tiles without breaking any of them. It is not required to follow the domino rules

Figure 1   The chessboard task

All the students were aware of the "domino rules" (as a prerequisite knowledge). The task was chosen because of its potentiality to be posed to either young or older students. At the same time different levels of reasoning can be applied for solving the problem. The answer is negative for both parts of the task. Each tile no matter its position in the chessboard must cover at the same time one white and one black square. In general k tiles placed on the chessboard will cover k black and k white squares. Yet, the chessboard we are dealing with has 32 black and 30 white squares. Thus, there exists an insurmountable problem. The chessboard can never be tiled. This solution could be accepted as a valid one even though it seems that it rests much on perception. Mamona Downs and Downs (2011) suggest a plausible mathematization of the solution but this demands the availability of mathematical tools that are available in tertiary education and an advanced mathematical background.

Students were asked to express written their line of thoughts so as to make clear their chain of arguments. They were also asked to answer why they chose (and insisted to) the specific line of reasoning. This question gave evidence about the existence of a certain belief. Their worksheets constituted our data. They were collected and coded according to (a) whether there could be tracked a certain belief that seemed to influence positively or negatively the student and (b) the distribution of these beliefs across grades. Then, in a second level, students' reasoning and according to its content was classified into two main types according to whether this reasoning could be considered valid or weak. Each student's effort was coded independently by two researchers. After that, they compared their results and whenever there was a disagreement they discussed it together looking for a convergence.

4 Results and Discussion

The task posed to the students was not procedural in the sense that it could be not solved by merely recalling certain algorithms and subsequently they did not have an idea how to proceed in solving the task. This is why they started by 'playing around' with possible paths from square A to B and it was during this process when they became able to decide what will be the answer and follow a certain line of reasoning based on a specific belief. The examination of the answers to the question concerning why they chose a specific approach, enabled us to distinguish three different beliefs that influenced the reasoning process of the participants. This was done on the basis of the form of the answers (*I think that...*, *My opinion is that...*, *I chose this approach because I believe...*, etc):

i. It is not necessary to use numbers, algorithms and operations (positive). (This is in accordance with Schoenfeld(1989): *Real math problems can be solved by common sense instead of the math rules you learn in school*)

ii. Considering many examples is in itself a proof (negative). (This is in accordance with Bieda, Holden and Knuth (2006) who claim that "*students believe that checking a few cases is sufficient*")

iii. Mathematical tasks should be solved only by using numbers and arithmetical relationships (negative). (This is in accordance with Tsamir and Tirosh (2002) who claim that students believe that "*when faced with a mathematical problem, a mathematical operation should be performed*").

All the students, no matter the belief they expressed, answered "correctly" that there is not solution for both parts of the task. But, depending on the belief they expressed, their reasoning was either valid or problematic. Below, we will show how these three beliefs made the students to succeed or fail in each part of the task.

Part 1

The students who expressed themselves through the first belief did not reject the perspective of using numbers but they were clear that this cannot be their unique option. This freedom allowed them to 'play around' without any bias towards certain approaches and this made them able to recognise the underlying structure yielded by the given situation. The majority of the convincing arguments were based on the usage of colours (Fig.2, left) which could be summarized by quoting a students' extract:

> S: Each domino covers simultaneously a black and white square on the chessboard. More precisely, we always start with a black square (A or B) and end with a white one. This is always happening no matter the path. The end of any path will be always a white square. However, I am asked to have a black square as the final one (A or B). This is impossible.

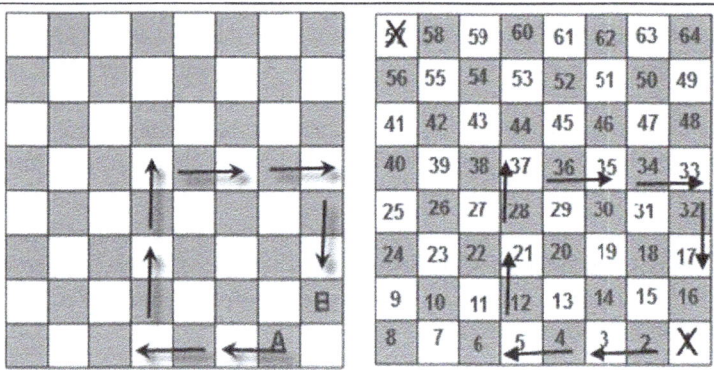

Figure 2   The usage of colours (left) and numbers (right)

Slightly different was another approach that was based on numbers (Fig.2, right). Even though the core idea of the argumentation is similar (even/odd and black/white), this one could be considered as more sophisticated (in the sense that numbers and their properties are involved), especially if one has in mind that this was proposed by a primary school student (grade 6). This student put consecutive numbers into the 64 squares from 1 to 64. The numbers must be follow a continuous line (right to left and then from left to right and so on).

Then he argued that by starting from A or B means that always the domino starts with an even and ends with an odd number no matter the path chosen. However, the asked final square (A or B) includes an even number. So, there could never be found any path satisfying the task's instruction.

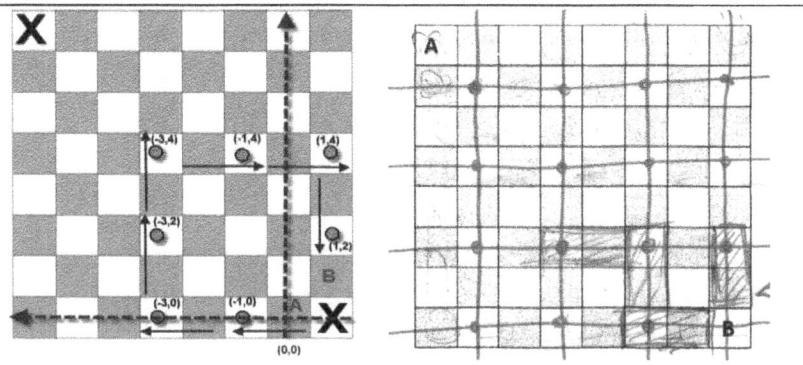

Figure 3 Using coordinate system (left) and intersection points (right)

It is worthwhile to mention here two other approaches presented by a Grade 11 pupil and a university student respectively. The former used a Cartesian coordinate system to show that it is not possible to find the asked path. He considered each square as a 'point' in a two-dimensional system of coordinates. The origin of the coordinate system (0,0) is square A and so square B would be defined by the pair (1,1). In Figure 3 (left) the final end of the dominos could be described as (-1,0) (-3, 0) (-3, 2) (-3, 4) (-1, 4) (1, 4) (1,2). According to his words:

> S: Starting from A (i.e., (0,0)) each domino ends to one of the following: (even for x-coordinate, even for y-coordinate), (even for x, odd for y) and (odd for x, even for y). The fact is that it is impossible to have a path ending to (odd for x-coordinate, odd for y-coordinate). However, the wanted square B is defined by the ordered pair (1,1) which means (odd,odd) and thus there is not such a path that ends to B.

In the latter one, a prospective elementary teacher drew straight lines that denoted the potential positions of the tiles in any path (Fig.3 right). The points of intersection of any two lines denote the end of the tiles. According to this, in the column that includes B there are only three intersection points where a tile can end. None of them is the wanted one. Thus, there is no solution to the problem. There were in total 9 correct solutions for the primary school students (9/81=11,11%), 4 for the secondary school students (4/35=11,42%), and 11 for university students (11/50=22%) (see Table 1).

Let's see now the working of the students who expressed themselves through the second and third beliefs. All of them decided that there cannot be any path from A to B, but their reasoning was not sufficiently convincing. Their argumentation was based on the surface features of the problem rather than the underlying structural elements, something that limited their problem solving abilities and made their reasoning to appear problematic. Thus, the students who were influenced by the second belief argued that every time they tried to find a path, one square was left over (i.e., B). For them, this was in itself a proof that the asked path does not exist.

Table 1 Arithmetical data for Part I.

	Correct	One square left over		Only 7 squares	
Primary	9/81(11.11%)	4 (Gr.5) 5 (Gr.6)	5/81(6.17%) 2 3	2/81(2.46%)	2 0
Secondary	4/35(11.42%)	3 (Gr. 10) 1 (Gr.11)	7/35(20%) 3 4	16/35(45.71%)	6 10
Tertiary	11/50(22%)		7/50 (14%)	14/50 (28%)	

The remaining ones (third belief) asked persistently for arithmetical relationships. Therefore, they argued that the row or the column with the missing square consists of 7 squares and so it is impossible to put integer number of dominos in this row or column because in that case 8 squares would be needed.

It is interesting to see how these two kinds of weak reasoning are distributed across the educational levels so as to make a comparison. Considering Table 1, one can see that as we are moving across the grades the students tend to lose their abilities to employ creative thinking. Therefore, they gradually become unable to work in an environment that does not offer a hint that would help them to make connections with known algorithmic procedures. The first problematic reasoning actually refers to the fact that the students tried several alternatives to obtain the desired path but without success. One square (i.e., B) is always left over and so it is not possible to find the asked path. This means that some examples were considered being a justification. But, the fact that they did realize a certain number of trials cannot guarantee that the task cannot be solved. Five primary school students (6.17%) adopted that approach. The corresponding percentages for secondary and tertiary (prospective elementary teachers) levels were 20% and 14% respectively. This is in itself an important finding. One could expect that the conclusions drawn from a limited number of trials would be generalized and accepted as a universal truth by primary school students. Obviously, their immaturity concerning mathematical thinking allows them to draw conclusions based just on specific cases. They do not feel the need to use a valid reasoning to convince themselves as well as their classmates or their teacher. What they see is enough for generalizing their conjectures or their conclusions. But it is not the same when the corresponding percentages for the other two educational levels are two or three times bigger than the primary school ones. This highlights an existing problem especially if we have in mind that our secondary students were oriented towards science and mathematics and the prospective elementary teachers would after a while be teaching in primary schools. The second problematic reasoning actually is based on an aspect of the parity "oddness vs. evenness". The main argument was that in the upper and lowest row of the chessboard was an odd number of squares (i.e., 7) and therefore it was not possible for the row to be completely covered by an integer number of unbroken tiles. All the 8 squares were needed to have the whole row covered. They did not think that there are possibly various paths and it was not required to cover completely the row in order to obtain the target path. The relevant arithmetical data for this second problematic reasoning were: (i) For primary education, 2 students (2.46%), (ii) for secondary education, 16 students (45.71%), and (iii) for tertiary education 13 prospective teachers (26%). The difference in the percentages between the very young students of primary school and the participants of the remaining two other levels of education is impressive and confirms the tendency of the older students to turn towards more formal ways of working such as numbers, operations and properties of numbers. This might be explained by considering that the negative impact of the same beliefs is stronger for older students rather than for younger ones.

Part 2

For the second part of the task the setting was different. The same imperfect chessboard had to be completely covered without following the rules that govern domino. Both the correct and the problematic reasoning of the students were developed around the same axes as it was in the previous part of the task: usage of colours and usage of numbers for the correct reasoning, and leftover squares and odd number of squares for the problematic reasoning. Actually, the revocation of the domino rules offered to the students more space to develop their reasoning.

The following extract is representative of how the students, who were influenced by the first belief, used the colours to explain why the chessboard cannot be covered completely by tiles:

S: Each tile, no matter of its placement in the chessboard, covers simultaneously a white and a black square. Given that two white squares are missing then two black ones are leftover. So, the black squares are more than the white and the two extra black ones cannot be covered since each one of them requires a white square to make a pair that would be covered by a tile.

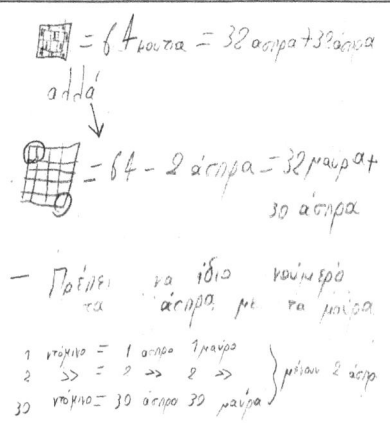

- 64 squares = 32 white + 32 black, but
- 64 – 2 white = 32 black + 30 white
- —We need equal number of white and black ones
- 1 tile = 1 white + 1 black
- 2 tiles = 2 white + 2 black
- 30 tiles = 30 white + 30 black

2 white needed

Figure 4   An effort for formalization in primary school

The second approach (expressed by a primary school student) used numbers to express (in essence) exactly the same reasoning as follows:

S: There are 32 black and 30 white squares in the chessboard. Each tile covers one black and one white square. Therefore, covering 30 white ones means that 30 black ones will be covered at the same time. Thus, 2 black squares remain that cannot be covered.

Then, the same student tried to express himself in a rather formal way (see Figure 4).

In total, eight primary students (9.87%), one secondary (2.85%) and six tertiary ones (12%) gave correct solutions (see Table 2). It seems that primary school students continued to keep their performance in more or less the same level compared to the rate of success in the first part. On the contrary, the percentages for the remaining educational levels were decreased to 2.85% (from 11.42%) and 12% (from 22%) for secondary and university students respectively.

Table 2 Arithmetical data for Part II

	Correct		Two squares left over		Only 7 squares	
Primary	8/81(9.87%)	3 (Gr.5) 5 (Gr.6)	15/81(18.51%)	7 8	7/81(8.64%)	2 5
Secondary	1/35(2.85%)	1 (Gr. 10) 0 (Gr.11)	8/35(22.85%)	1 7	6/35(17.14%)	3 3
Tertiary	6/50(12%)		14/50 (28%)		15/50 (30%)	

Again, it is interesting to see how the students were distributed between the two beliefs that inspired the two kinds of problematic reasoning. On the one hand there were students who made several attempts to cover the whole chessboard without success since "always two squares are leftover". On the other hand, there were students who decided that it was not possible to cover the chessboard and this decision was based on the fact that the upper and lowest row consisted of an odd number of squares whereas an even number is needed in order to completely cover the row. The data in Table 2 confirm the general impression drawn from Table 1 that the impact of the same mathematical beliefs becomes stronger as we move from younger (primary school) to older students (secondary and tertiary education).

5 Conclusions

Research has showed that beliefs can support or mislead the solver no matter his/her age. Using this as a starting point this paper tries to explore whether there is uniformity in the extent these beliefs affect the students across the educational levels. The findings show that in primary school the positive impact of beliefs leaves space for students to employ creative reasoning during non-routine problem solving. However, as far as the negative impact of beliefs is concerned, moving from primary education to secondary and (in some cases) to tertiary education it seems that the students, trapped in their beliefs, mainly seek algorithmic strategies and connections to formal mathematics for solving such tasks (see also Garofalo and Lester, 1985). This is confirmed by the tendency of many of our students to employ odd and even numbers in order to reach an answer. Moreover, the low percentages of successful approaches for the secondary students or the prospective elementary teachers show that they are unable to respond in problems that cannot be placed under a certain type of "well-known" ones. On the contrary it is interesting to see primary school students to perform almost with the same rate of success as the older students to the same problems and at the same time to realize that these young students obtained the lowest scores in the percentages concerning the problematic kinds of reasoning which means that the same beliefs are less determinant for younger students (who are rather less preoccupied against certain ways of approaching problem solving). A potential explanation might be the way our educational system itself works. During primary school years there is no pressure to obtain certain scores in exams and thus the students are offered the opportunity to develop creative thinking or to exploit their imagination. However, as we are moving towards secondary education the whole system becomes focused on the national exams which determine whether the students will be accepted for college studies. This means that they must be equipped with very concrete algorithmic techniques that will enable them to solve certain types of problems. However, it seems that very often this way of educating pupils is accompanied by the establishment of certain beliefs about how one can approach problem solving in mathematics. Unavoidably, this emasculates the students' ability to work in a less formal but more creative level and leaves them with a fault impression about what mathematics problems really are and how one can cope with them. Therefore, it is needed to give emphasis to the development of this ability from the early schooling by giving to the young students very often opportunities to be involved in situations that challenges their beliefs and provoke their imagination and the production of logical arguments to obtain a truth status.

6 References

Ball, D. Joyles, C., Jahnke, H., & Movshovitz-Hadar, N. (2002) The Teaching of Proof. In L.I. Tatsien (Ed.), *Proccedings of the International Congress of Mathematicians,* (v.II, pp. 907-920). Beijing: Higher Education Press.

Bieda, K., Holden, C., & Knuth, E. (2006). Does proof prove? Students' emerging beliefs about generality and proof in middle school. In *Proceedings of the 28th Annual Meeting of the North America Chapter of PME,* (v.2, pp. 395-402).

Carlson, M. (1999). The mathematical behaviour of six successful mathematics graduate students: Influences leading to mathematical success. *Educational Studies in Mathematics, 40*(3), 331-345.

Crawford, K., Gordon, S., Nicholas, I. & Prosser, M. (1994). Conceptions of mathematics and how it is learned: The perspectives of students entering university. *Learning and Instruction, 4,* 331-345.

English, L.D. (1997) Intervention in children's deductive reasoning with indeterminate problems. *Contemporary educational psychology, 22,* 339-362.

Fosnot, C., & Jacob, B. (2010). *Young mathematicians at work.* NCTM.

Garofalo, J., & Lester, F.K. (1985). Metacognition, Cognitive Monitoring and Mathematical Performance. *Journal for Research in Mathematics Education, 16*(3), 163-176.

Hanna, G. (2000). Proof, explanation and exploration: an overview. *Educational Studies in Mathematics, 44*(1), 5-23.

Lithner, J. (2008). A research framework for creative and imitative reasoning. *Educational Studies in Mathematics, 67*(3), 255-276.

Maher, C., & Martino, A. (1996). The development of the idea of mathematical proof: A 5-year case study. *Journal for Research in Mathematics Education, 27*, 194-214.

Mamona-Downs, J. & Downs, M. (2011). Proof: a game for pedants? *Proceedings of CERME 7,* 213-222, Rzeszow, Poland.

Mamona-Downs, J. & Downs, M. (2013). Problem Solving and its elements in forming Proof, *The Mathematics Enthusiast, 10*(1-2), 137-162.

Schoenfeld, A. (1985). *Mathematical problem solving.* Orlando: Academic.

Schoenfeld, A. (1992). Learning to think mathematically: Problem solving, metacognition and sense-making in mathematics. In D. Grouws (Ed.), *Handbook of Research in Mathematics Teaching and Learning* (pp. 334-370). New York: Macmillan P. C.

Schoenfeld, A. (1989). Exploration of Students' Mathematical Beliefs and behaviour. *Journal for Research in mathematics Education, 20*(4), 338-355.

Sumpter, L. (2013). Themes and interplay of beliefs in mathematical reasoning. *International Journal of Science and Mathematics Education* (to appear).

Tsamir, P., & Tirosh, D. (2002). Intuitive Beliefs, Formal definitions and Undefined Operations: Cases of Division by Zero. In G.C. Leder, E. Pehkonen, and G. Torner (Eds.), *Beliefs: A Hidden variable in Mathematics education?* (pp. 331-344).

Investigating Mathematical Beliefs by Using a Framework from the History of Mathematics

Lenni Haapasalo, Bernd Zimmermann

University of Eastern Finland, Finland; Friedrich Schiller University of Jena, Germany

Content

1 Introduction 198
2 Theoretical Background..198
3 Aims and Methods..202
    3.1   The study among student teachers 202
    3.2   The study among pupils 202
4 Results 203
    4.1   Comparison of the beliefs among Finnish and German student teachers...203
    4.2   A pick-up from the study among pupils 205
5 Discussion 207
6 References 208

Abstract

In the theoretical part of the article a framework of eight activities and motives is sketched (calculate, apply, construct, argue, order, find, play, evaluate), which proved to be successful along the history of mathematics. Furthermore, some arguments are presented and discussed why this framework is useful for studying mathematical beliefs. The empirical part is about two case studies using this network carried out in Joensuu (FIN) and Jena (GER). The goal of the first study was to compare mathematical beliefs of student mathematics teachers in Finland and Germany. The second part is about the influence of using a handheld calculator on the belief of a pupil. The first study reveals that neither in Finland nor in Germany the school mathematics seems to give much support for these activities, in Finland university mathematics even less. The only exception is calculating, for which the both institutions seem to give overdose. On the other hand, the finding of the second study that voluntary playing with progressive technology, even during a short period of time, might shift mathematical beliefs in a positive way.

1 Introduction

We start by making some remarks about mathematical beliefs before we sketch a quite new framework drawn from history of mathematics and argue, why we think, that such additional approach is useful. It encompasses eight main motives and activities, which proved to lead frequently to new mathematical results at different times and in different cultures for more than 5000 years. In the empirical part we present two case studies based on questionnaires developed by the first author on the background of the aforementioned activities that we will call Z-activities, shortly. Finally, we present some results, a discussion some hypotheses.

2 Theoretical Background

We used the following preliminary characterization of mathematical belief:
Mathematical beliefs are views of mathematics and mathematics instruction, which have cognitive (Skemp 1979, Schoenfeld 1985, Bogdan 1986), affective (Frank 1985, Schoenfeld 1985, Yackel/Carter 1989) and normative (Zimmermann 1991) aspects, which influence learning and teaching of mathematics.

The main emphasis of this part will be placed on the first and last aspect. We present here the origin of this (completely new) approach, which seems to be necessary to better understand its motivation and to relate it to some original research of the MAVI-group.

Our starting point is the habilitation of Zimmermann 1991, consisting of two parts: (1) mathematical beliefs of teachers and pupils, and (2) history of mathematical problem solving (esp. heuristics). The original goal of the first part was to make an empirical contribution to the attempt to develop a meta-theory of mathematics education (cf. Zimmermann 1979). The theoretical background of the corresponding teacher-questionnaire (included in Zimmermann 1991) had been constituted by studies presented in Zimmermann 1981, 1983, 1987. Questionnaires had been administered in 1988 to teachers (n=110) and pupils (n=2600, mainly grade 7) in Hamburg and some suburbs. Some outcomes of these studies had been published in Zimmermann 1991, 1997, 1997a, 2002. Later on, the questionnaires were used as tools to detect possible points of departure of teachers' and pupils' beliefs in a project about open problem solving, which was planned already 1987 in Germany (cf. Kießwetter/Zimmermann 1987), but finally could be conducted in Finland only (cf. e.g. Pehkonen/Zimmermann 1990). The questionnaires were translated by Pehkonen and administered in several countries (Finland, Sweden, Hungary, the US and Estonia, cf. e.g. Pehkonen 1994). Hannula et al. 2013 used it as one tool to compare Baltic States and to detected possible changes with respect to teacher-beliefs.

There was some side effect of the second part of this habilitation (about history of problem solving) in studying mathematical beliefs. This long-term study of history of mathematics (encompassing some 1000 text-resources), revealed the following eight main sustainable motives and activities, which proved to lead frequently to new mathematical results at different times and in different cultures for more than 5000 years. These different activities are connected and interrelated in many ways, which is represented in Fig. 1 by the connecting lines.

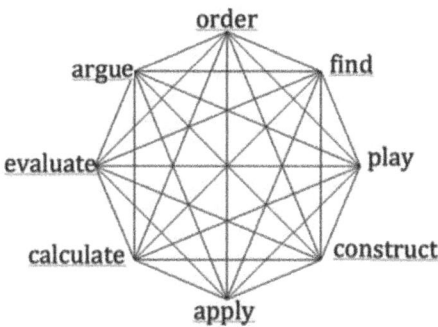

Fig. 1   Z-activities which proved to be successful in producing new mathematics (Zimmermann 2003, p. 42).

To get at least some impression about the meaning of the words from Fig. 1, we give here a short summary: At the very beginning of nearly all large cultures the documentation of quantities and their manipulation was of major interest (from the "Ishango-bone" up to scientific computing of today). This lead to the first important mathematical activity: *calculate*. Problems, e. g., from astronomy and agriculture, are until our days - from every-day problems up to space-industry and ecology - very important domains to *apply* mathematics and to develop new mathematical models, respectively (cf. the dominating "philosophy" of PISA). *Construct* is the most important activity, not only in (classical) geometry but also in architecture – which was taken as a part of mathematics for a long time.

These three activities are to some extent the oldest ones and therefore they form the basis of Fig. 1.

Methods to *invent* something like heuristics (e.g., working backwards, analogizing, successive approximation, change of representation etc.) were applied at least implicitly since the ancient times of, e. g., the Babylonians and Chinese. To *argue*, esp. proving is at the core of modern mathematics and belongs to the more challenging mathematical activities. The tension to bring new knowledge, a set of new theorems or clusters of solved problems in a systematic *order*, including approaches to axiomatization (Euclid), led very often to a deeper understanding and to more insight into theoretical interrelations.

These activities are more challenging and sophisticated than those at the bottom of Fig. 1. Therefore we put them at the top.

Finally, there are two activities, which seem to be neglected to some extent, but which proved to be major stimulators over and over again, too.

Striving for religious cognition as well as for esthetical (geometric or proof-) configurations and related systems of values generated also new problems and their solutions and produced in this way also new mathematical knowledge during history of mathematics (e.g., Vedic geometry, combinatoric, tessellation and ornaments in Islamic Mathematics). So the underlying activity of *evaluate* proved to be of importance, too.

The same holds for an approach to mathematics by *play*ing and the development of recreational mathematics. In this way very oft new branches of mathematics were created like stochastic and game-theory.

We skip a more detailed discussion on the development of the Z-activities because it can be found in Zimmermann (1999, 2003). Some first results coming from this approach were already presented at the MAVI-meeting in Duisburg (Zimmermann 1998).

There have been carried out several studies, which detected general phases in history (Spengler 1988), general styles of thinking in the history of science (Crombie 1994) and general laws in history of mathematics (Wilder 1981). None of them centered on such activities, which produced new mathematics over a long range of time.

If we look back on these activities from an educational point of view, we realize that they are not at all less important for today's mathematics instruction, especially if creative activities of pupils are stressed, too (cf. OH 2004, p. 17,

105). The interconnections between these activities, represented in Fig. 1, correspond to the general goal (of learning) to achieve a high degree of flexibility in thinking, to foster connecting thinking and mastering many routine activities as well.

So we have already one good reason, why we took history of mathematics as an additional framework. Before we outline some more reasons, we make some remarks about mathematics education of today.

Trends in mathematics education are subjected to change. At the moment constructivism, real-life mathematics, use of computers and "assessment" are "in". Some possible pitfalls are as follows:

During the last years there can be observed a slowly decrease of the importance of the mathematical content in some mathematics education communities. Some hints (no "proofs", of course) are:

- A colleague from Hungary, who is teaching mathematics education in Sweden since 1989, said: "In Hungary mathematics is taught, in Sweden nomathematics is taught". For us this highlights to some extent the conflict between the very excellent Hungarian tradition in mathematics, neglecting very often "real life" and overloading the average pupil, and the tendency in the West, to concentrate sometimes too much on activities, where instant motivation, communication for a goal in itself and the immediate use for tomorrow - sometime on expense of mathematical content - is to be achieved.

- This opinion was also expressed by Prof. Kaenders (Cologne), when he criticized Prof. J. de Lange after his lecture "Problems with problem oriented mathematics instruction", by saying, that the Dutch approach of "realistic mathematics" includes the danger to diminish the mathematical content and understanding.

- At least in Germany more and more the assessment-tail seems to "wag" the content dog. One possible indicator: There had been introduced a special slot "mathematics in mathematics education" at the last GDM-conference in Münster 2013. By systematic analysis of scientific papers of last decades, published in the Journal of Mathematics Education, Jahnke comes to the conclusion, that the content is "slowly vanishing out of mathematics education" (Jahnke 2010).

Keeping these problems in our mind, the reference to history of mathematics has the following additional advantages:

- It can be taken as a mainly (but not only) cognitive long-term-study about most productive activities, as we did.
- Results could be more independent from fashion-waves and more sustainable.

- Results can be taken not only as an additional framework for measuring the outcome of (as we soon will see) but also for orientation in mathematics instruction (we did so by orienting the development of a new textbook-series on this framework, cf. Cukrowicz/Zimmermann 2000).

3 Aims and Methods

It was the idea of the first author, to use the framework of the previous section for studying mathematical beliefs of student teachers and pupils. He was especially interested to what extent use of ICT changed these beliefs (cf. e.g. Haapasalo 2007, Haapasalo/Eronen 2011, Haapasalo/Hvorecky, J. 2011, Haapasalo/Samuels, Haapasalo/Zimmermann 2011, Haapasalo/Eskelinen 2013). Based on the "octagon" in Fig. 1, he developed several Likert-scaled questionnaires (scale 1-5 or -2 to +2) to measure different types of so-called profiles of student mathematics teachers and pupils.

3.1 The study among student teachers

The student teachers were asked to answer the following questions:

a) How good do you think you are in performing each activity of the octagon (self-confidence of student teachers)?

b) To what extent do you think your received support by school mathematics to carry out these activities (support by school-mathematics)?

c) To what extent do you think you received support by university mathematics to carry out these activities (support by university mathematics)?

The corresponding questionnaires were administered to 25 student teachers at the University of Eastern Finland in Joensuu and to 20 student teachers (Gymnasium) at the Friedrich-Schiller-University of Jena. For the comparison the Likert-scale was changed from 1 to 5 into -2 to +2).

3.2 The study among pupils

The pupils in Joensuu were asked by their teacher (Lasse Eronen) to cope with linear functions in their summer holidays with the opportunity to use ClassPad (and to write a protocol). Before (in May) and after (in August) these activities they were asked to answer the following questions (from 1= scant or not at all until 5= very strong):

a. How strong does each activity of the octagon appear when using the term 'mathematics' (view of mathematics)?

b. How good the person does think he or she is performing each activity of the octagon (self-confidence, cf. Finish Curriculum 2004[1])?

c. How strong a computer does give support to each activity of the octagon (influence of computers)?

Furthermore, some interviews were conducted to validate the questionnaire and to get some deeper information about the possible changes of the opinions.

4 Results

4.1 Comparison of the beliefs among Finnish and German student teachers

Self-confidence

Fig. 2 illustrates that the self-confidence among Finnish students seems to be a bit higher than among German ones. The latter feel mainly stronger in "find" and in "argue" than the Finnish ones. Evidently the Finnish students trust more on their creative abilities and abilities to argue. This might be related to the different history and traditions in both areas during the last 80 years.

---

[1] In Finish schools "self-confidence and responsibility for one's own learning" are the first learning- goals to be achieved in mathematics with beginning of grade 6 (cf. "...luottamaan itseensä ja ottamaan vastuun omasta op-pimisestaan matematiikassa...", OH 2004, p. 110).

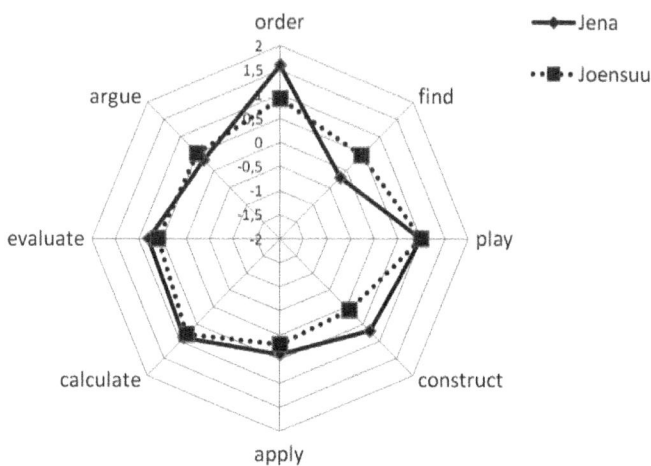

Fig. 2 Self-confidence of student teachers Jena – Joensuu

Support by school mathematics

It is only "calculate" that got support from objects' former mathematics instruction. Maybe the most interesting outcome is that the German subjects found essentially less support for "find" than their Finish colleagues, whilst for "play" the outcome is opposite.

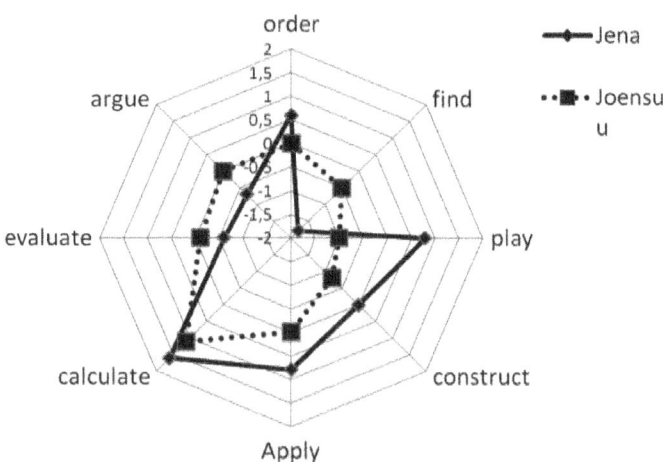

Fig. 3 Support by school-mathematics of student teachers in Jena and Joensuu

Support by university mathematics

The overwhelming impression of the German subjects seems to be, that they are mainly supported in calculate still at University. Nearly no support they feel to get - once again - for "find", which might reinforce the impression that creativity was outside the scope of their experience. The Finnish subjects experienced university mathematics offered nearly the same modest support for all activities.

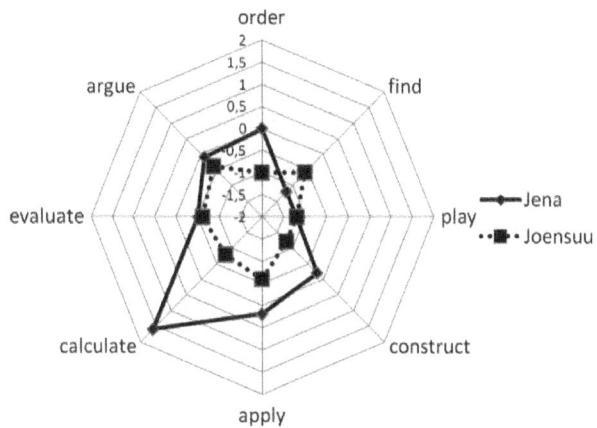

Fig. 4 Support by university-mathematics of student teachers in Jena and Joensuu

4.2 A pick-up from the study among pupils

At the beginning of the so-called ClassPad project, this unfamiliar tool was demonstrated briefly to 8th grade students to give them opportunity to play with it voluntarily during their summer holiday with concepts of 9th grade mathematics (such as a linear function). Their only required duty was to write some kind of portfolio if they worked with the tool. Because a comprehensive analysis has been done in Eskelinen & Haapasalo (2010), we present here the results of one pupil who is called Susi.

View of Mathematics

Fig. 5 represents Susi's answers with respect to the importance of these activities to do mathematics. While Susi's picture of mathematics was dominated at the beginning by the applicability Mathematics, after the holidays and coping with linear functions, this aspect was shifted more into the background.

Arguing and methods of finding became more important. An interview in August revealed: *"Now I know better and see things in different light".*

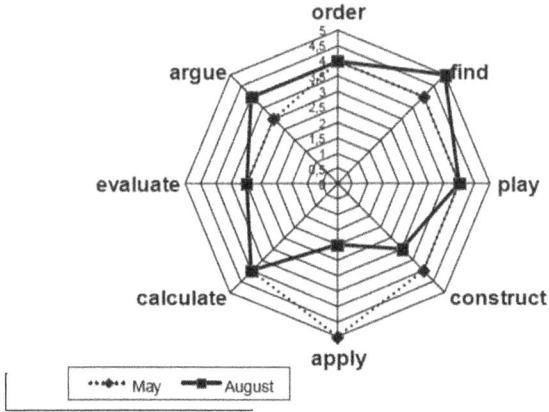

Fig. 5   Susi's view of mathematics before and after the ClassPad work.

Self-confidence to make mathematics:

Once again, after the period of learning in the summer-holidays, Susi had less confidence in applying mathematics. This might be due to the fact that the guidelines of the teacher for that time did not focus on applications of linear functions. They were treated later in the classroom. After using the ClassPad, Susi demonstrated increasing self-confidence in being able to "order" and to "find".

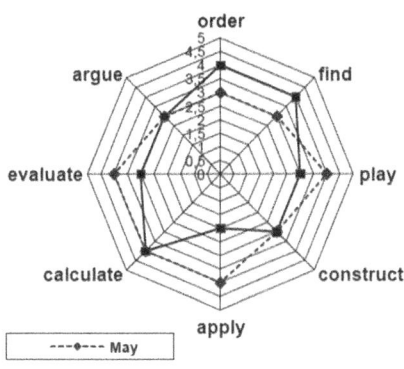

Fig. 6   Susi's view of computer's role in making mathematics

Impact of computers on mathematical belief

The clearest increase of agreement was in "play". This was underpinned by the following statement in an interview:

"*In May I could not even think to play ClassPad in summer holiday. However, I noticed, that it was very capable for playing with mathematics.*"

Some decrease of agreement could be observed as well: "Calculate", "order" and "find" were less stressed.

Once again, Susi's remark in an interview sheds some light on this observation: "*ClassPad is suitable for calculating, but if you want to learn how to calculate, you have to do something by hand*".

The protocols of Susi offered some more insight how she discovered for herself by appropriate use of ClassPad - sometimes in the middle of the night - concepts as slope and different equations to represent linear functions.

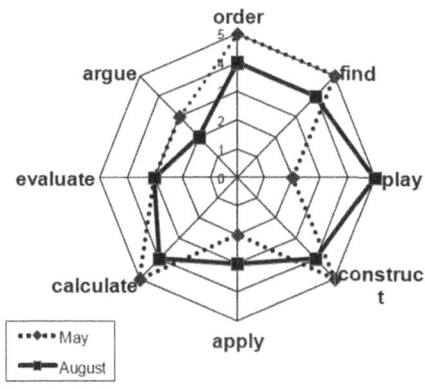

Fig. 7 Susi's view of computer's role in making mathematics

5 Discussion

Our framework to analyze mathematical beliefs incorporates not only modern ways of understanding, but it is much more encompassing. To facilitate the understanding of this new approach - we made some remarks about is origin and its relation to the MAVI-community. As it is typical for exploratory studies, we wanted to get some ideas and possible hypotheses, which could be followed in a more systematic way later on. This approach seems reasonable for us because of the novelty of the whole approach.

As to the comparison of student teachers from Germany and Finland, we derived from our specific observations the following hypotheses (based on the aforementioned framework):

H1. Finnish student teachers are less self-confident than German students.

H2. German students (at least from Eastern Germany) seem to feel less creative in mathematics and experienced less support at school and university in this respect.

H3. German student teachers (at least from Eastern Germany) feel stronger to order there mathematical work and get more support in playing and calculating at school and university in relation to their Finish colleagues.

H4. There might be still some difference between the beliefs of western and eastern German teachers.

H5. The differences between Western and Eastern German student teachers are smaller than between corresponding teachers.

In our case study with Susi we observed, that she demonstrated in the given voluntary and free context several positive learning gains. This observation triggers at least the following question: To what extent (for whom?) this experience can be generalized (with respect , e. g., to age of the students, quality of the teacher-"coach", social setting, system of values for teaching and learning)?

Ongoing research of Prof. Haapasalo probably reveals more interesting phenomena and hypotheses, not only regarding math instruction itself but more generally: Where does the self-confidence in doing Z-activities actually come from (cf. Haapasalo & Eskelinen 2013).

6 References

Bogdan, R. J. (ed. 1986). *Belief. Form, Content and Function*. Oxford: Clarendon Press.

Crombie, A. C. 1994. *Styles of Scientific Thinking in the European Tradition: The History of Argument and Explanation Especially in the Mathematical and Biomedical Sciences and Arts*. 3 Volumes. London: Duckworth.

Cukrowicz, J.; Zimmermann, B. 2000. *MatheNetz Ausgabe N; Klasse 7-11; Gymnasien*. Braunschweig: Westermann.

Eronen, L., Haapasalo, L. 2010. Making Mathematics through Progressive Technology. In: B. Sriraman, C. Bergsten, S. Goodchild, G. Palsdottir. B. Dahl and L. Haapasalo (Eds.), *The First Sourcebook on Nordic Research in Mathematics Education*. Charlotte, NC: Information Age Publishing, 701-710.

Frank, M. L. 1985. *Mathematical Beliefs and Problem Solving*. Diss. Purdue Univ., UMI 8606543.

Haapasalo, L. 2007. Adapting Mathematics Education to the Needs of ICT. *The Electronic Journal of Mathematics and Technology*, 1 (1), 1-10.Internet: https://php.radford.edu/~ejmt/deliveryBoy.php?paper=eJMT_v1n1p1 .

Haapasalo L.& Eronen L. 2011. Looking Back and Forward on the Light of Survey Studies Related to Mathematics Teacher Education. In: H. Silfverberg & J. Joutsenlahti (Eds.), *Integrating Research into Mathematics and Science Education in the 2010s. Proceedings of Annual Symposium of the Finnish Mathematics and Science Education Research Association*, 67-84.

Haapasalo, L. & Eskelinen, P. 2013. Elementary level trainee teachers' views of teaching mathematics and the usage of technology at the beginning of their didactical courses. In M. Hähkiöniemi, H. Leppäaho, P. Nieminen & J. Viiri (eds.), *Proceedings of the annual conference of Finnish Mathematics and Science Education Research Association* (pp. 25-33). University of Jyväskylä, Department of Teacher Education, Research report ••xx. Jyväskylä: University Press.

Haapasalo, L. & Hvorecky, J. 2011. Evaluating the Zimmermann Octagon within Research Standards. In: T. Fritzlar, L. Haapasalo, F. Heinrich & H. Rehlich (eds.). *Konstruktionsprozesse und Mathematikunterricht*. Hildesheim: Franzbecker, 145-152.

Haapasalo, L. & Samuels, P. 2011. Responding to the Challenges of Instrumental Orchestration through Physical and Virtual Robotics. *Computers & Education* 57 (2), 1484-1492. Internet: http://www.sciencedirect.com/science/article/pii/S0360131511000443

Haapasalo, L. & Zimmermann, B. 2011. Redefining school as pit stop: It is the free time that counts. In: W-C Yang, M. Majewski, T. de Avis & E. Karakirik (Eds.) *Integration of technology into mathematics education: past, present and future*. Proceedings of the Sixteenth Asian technology Conference in Mathematics, 19-23 September in Bolu, Turkey, 133-150.

Hannula, M. S., Pipere, A., Lepik, M., Kislenko K. 2013. Mathematics Teachers' Beliefs and Schools' Micro-Culture as Predictor of constructivist Practices in Estonia, Latvia and Finland. In: Lindmeier, A. M. & Heinze, A. (Eds.). *Proceedings of the 37th Conference of the International Group for the Psychology of Mathematics Education*, Vol. 2, pp. 433-440. Kiel, Germany: PME.

Jahnke, Th. 2010. Vom mählichen Verschwinden des Fachs aus der Mathe- matikdidaktik. (About the slowly vanishing of the subject out of mathematics education). In: Lindmeier, A., Ufer, S. *Beiträge zum Mathematikunterricht 2010*. München, 441-444.

Kießwetter, K., Zimmermann, B. 1987. Modellierung und Initiierung von Lernprozessen bei der selbständigen Betätigung von Schülern in (elementar-) mathematischen Problemfeldern. (Modelling and initiating of learning processes involved with independent activities of pupils in (elementary) mathematical problemfields. Research-application, adopted and conducted in Finland by Erkki Pehkonen after acceptation by the Finish Academy of Science. Cf. Pehkonen/Zimmermann 1990.

Opetushallitus (Finish School-board) 2004. Perusopetuksen opetussuunitelman perusteet 2004 (Fundamental guidelines of the Finish Curriculum for the instruction at the comprehensive schools).

Pehkonen, E., Zimmermann, B. 1990. Probleemakentät Matematiikkan Opetuksessa ja niiden Yhtetys Opetuksen ja Oppilaiden Motivaation Kehittämisen. Osa 1: Teoreettinen tausta ja tutkimus-asetelma. (Problemfields in Mathematics Instruction and their Relation to the Development of Instruction and Pupils' Motivation. Part I. Theoretical Framework and Research Objectives). Helsinki: Helsingin yliopiston oppetajankoulutuslaitos, Tutkimuksia 86, 105.

Pehkonen, E., & Lepmann, L. 1994. Teachers' conceptions about mathematics teaching in comparison (Estonia - Finland). In: M. Ahtee & E. Pehkonen (Eds.), *Constructivist viewpoints for school teaching and learning in mathematics and science*. Helsinki: University of Helsinki, 105-110.

Schoenfeld, A. H. 1985. *Mathematical Problem Solving*. Orlando: Academic Press.

Skemp, R. R. 1979. *Intelligence, Learning, Action*. Chichester: Wiley.

Spengler, O. 1988. Der Untergang des Abendlandes. (The Decline of the West). 9th printing. München: DTV 838.

Thompson, A. G. 1985: Teacher's Conceptions of Mathematics and the Teaching of Problem Solving. In: Silver, E. A. (ed.): *Teaching and Learning Mathematical Problem Solving: Multiple Research Perspectives*. Hillsdale: Lawrence Erlbaum Associates, 281 - 294.

Thompson, A. G. 1992. Teacher's Beliefs and Conceptions: A Synthesis of the Research. In: Grouws, D. A. (ed.): *Handbook of Research on Mathematics Teaching and Learning. A Project of the NCTM*. New York: Macmillan Publishing Company, 127 - 146.

Wilder, R. L. 1981. *Mathematics as a Cultural System*. Oxford: Pergamon Press.

Yackel, E.; Carter, C. S. 1989. *Beliefs, Emotional Acts and Mathematical Activity: A Case Study of Change*. Poster presented at the 13th Conference of the International Group for the Psychology of Mathematics Education, Paris, 244-251.

Zimmermann, B. 1979. Einige Vorbemerkungen zu einer Metatheorie der Mathematikdidaktik. In: *Beiträge zum Mathematikunterricht 1979*. Hannover: Schroedel.

Zimmermann, B. 1981. Versuch einer Analyse von Strömungen in der Mathematikdidaktik. In: *ZDM* 1/1981, 44 - 53.

Zimmermann, B. 1983. On Some Trends in German Mathematics Education ERIC, ED 221 372, Ohio: Columbus.

Zimmermann. B. 1987. Suuntauksia Saksalaisessa Matematiikkassa (About some Trends in German Mathematics Education). In: *Dimensio* 8/1987, 10 - 12.

Zimmermann, B. 1990. Heuristische Strategien in der Geschichte der Mathematik. In: M. Glatfeld, M. (Ed.). *Finden, Erfinden, Lernen. Zum Umgang mit Mathematik unter heuristischem Aspekt.* Frankfurt a. M.: Peter Lang, 130 - 164.

Zimmermann, B. 1991. *Heuristik als ein Element mathematischer Denk- und Lernprozesse. Fallstudien zur Stellung mathematischer Heuristik im Bild von Mathematik bei Lehrern und Schülern sowie in der Geschichte der Mathematik.* University of Hamburg: Habilitation. Internet: http://users.minet.uni-jena.de/~bezi/Literatur/ZimmermannHabil-beliefs-historyofheuristicsinclABSTRACT.pdf.

Zimmermann, B. 1997.Vorstellungen über Mathematik und Mathematikunter- richt bei Lehrerinnen und Lehrern. (Teachers' Conceptions about Mathematics and Mathematics Instruction.) In: *Beiträge zum Mathematikunterricht*, 576-579.

Zimmermann, B. 1997a. On a Study of Teacher Conception of Mathematics Instruction and some Relations to TIMSS. In: Törner, G. (ed.): *Current State of Research on Mathematical Beliefs.* Proceedings of the MAVI-Workshop University of Duisburg, April 11-14, 1997, p. 117 - 126.

Zimmermann, B. 1998. On Changing Patterns in the History of Mathematical Beliefs. In: G. Törner (Ed.): *Current State of Research on Mathematical Beliefs VI.* Proceedings of the MAVI-Workshop University of Duisburg, 107-117.

Zimmermann, B. 1999. Kreativität in der Geschichte der Mathematik (Creativity in the history of Mathematics). In: B. Zimmermann, G. David, T. Fritzlar, and M. Schmitz (Eds.) *Kreatives Denken und Innovationen in mathematischen Wissenschaften. Jenaer Schriften zur Mathematik und Informatik.* Math/Inf/99/29, 227 – 245. Internet: http://users.minet.uni-jena.de/~schmitzm/kreativesdenken/tagband/zimmermann/zimmermann.pdf

Zimmermann, B. 2002. Vorstellungen über Mathematik und Mathematikunter- richt von Lehrerinnen und Lehrern verschiedener Schularten. (Conceptions about Mathematics and Mathematics Instruction from Teachers of different School Types.) In: *Der Mathematikunterricht.* Heft 4/5 2002, 7 - 25

Zimmermann, B. 2003. On the Genesis of Mathematics and Mathematical Thinking - a Network of Motives and Activities Drawn from the History of Mathematics. In L. Haapasalo & K Sormunen (Eds.) *Towards Meaningful Mathematics and Science Education.* Bulletins of the Faculty of Education 86, 29-47.

Students' Gazes: New Insights into Student Interactions

Peter Liljedahl, Chiara Andrà

Simon Fraser University (Canada) and University of Torino (Italy)

Content

1 Methodology 214
   1.1   The Problem 214
   1.2   The Students 215
   1.3   The Data 215
2 Interactive Flowchart 216
3 Interactive Flowchart + Gaze 219
4 Fictional Writing 221
5 Discussion 225
6 References 225

Group interactions are socially and affectively charged environments wherein these dimensions cannot be separated out from the learning that takes place. To this end, we examine the cognition-emotion interplay visible in a video clip of four students working together to solve a problem. By resorting to different analytic tools crafted for capturing communicational, cognitive, and affective aspects of an interaction we come to delineate a frame which not only integrates different methodologies, but also allows us to unfold the students' inner word through writing fictitious, but likely, inner speeches. We discuss the reciprocal role played by tools that allow us to access the students' interpersonal and psychological worlds.

The true mystery of the world is the visible, not the invisible.

Oscar Wilde

Abstract

Group interaction is gaining more and more prominence in curricula around the world, for many reasons ranging from teaching to acknowledging the value of dialogue as scaffolded metacognition (Sfard, 2001). But group interactions are complex socially and affectively charged environments wherein affect cannot be separated out from learning. Roth & Radford (2011) argue that we need to overcome the dualistic approach between the individual's interior space and his social interaction, and focus more on sociocultural conditions. Learning occurs in and through relations with others driven by collectively motivated activity. Activity is a process with inner contradictions, differentiations, transformations, as well as emotions—necessary for the activity and responsible of its development. In this article we explore some of these processes through the analysis of a group interaction. More specifically, we explore the role of emotions, motivation, will, as well as knowledge and learning visible, and invisible, within the interaction.

1 Methodology

At the core of the research presented here is a 45 second video clip of a group of four students working on a mathematics problem.

1.1 The Problem

The problem was inspired by the work of Iversen and Nilsson (2005), who used a similar task to see how students make sense of random phenomena. The problem is:

> A robot walks along a corridor, it turns right with probability 1/3 and it turns left with probability 2/3. The map shows the labyrinth where the robot has to move. Compute the probability for the robot to be in each of the rooms.

The problem was crafted so as to use the representation provided by the task in order to introduce the concepts and the algorithms related to the tree diagram: why should one multiply subsequent branches? Why and when should one add? The task was presented like a game, and the students seemed willing to work on it as such.

1.2 The Students

The task was used as part of a series of four lessons on probability in a grade 9 (14-15 year olds) class in Bologna, Italy. The task formed a significant portion of the second lesson. The school is located in a low socio-economic neighborhood and all the students at the school live in this neighborhood. Four students (Luca, Fabio, Davide, and Marco) were selected to be videotaped while they worked on the task as a group.

1.3 The Data

These four students worked on the task in a separate room and were filmed by a grade 12 student from the same school. The entire session lasted 50 minutes. The first 5 minutes of this video were transcribed. From this the first 45 seconds were selected to constitute the data for the research being presented here (see table 1). This subset of the data was selected because it exemplified some very interesting and turbulent undercurrents of group interactions.

Table 1: Transcription of First 45 Seconds of Video

00:00	M:	To the left two thirds, to the right one third.
00:01	D:	Yes, I don't remember. (*speaks over M*)
00:03	M:	Then it goes two thirds, two thirds.
00:06	M:	Can you give me a pen, please?
00:07	L:	No, let's do the first case, which is the one where it goes always ...
00:10	M:	... left. You have two thirds here ...
00:11	L:	That is the most probable one. (*speaks over M*)
00:13	M:	...and here is one third.
00:15	L:	Should you erase?
00:16	M:	Yes, bravo!
00:17	D:	I'm cute!
00:19	M:	Two thirds and here one third, hence these two thirds...
00:21	F:	... they g ... they go ....
00:22	M:	Two thirds of two thirds.
00:25	D:	But ... but what are you saying? Then no ...
00:27	M:	Of these two thirds you should do ...
00:28	D:	We have ... but what do we have to compute? (*speaks over M*)
00:30	L&M:	The probability that the robot will arrive in each one ...
00:34	M:	of these rooms.
00:35	D:	In the meantime, let's see ...
00:36	L:	Why don't we first compute how many probabilities there are in all?
00:37	M:	To me this is the room with the highest probability.
	D:	Why?
00:42	L:	There are 8 in all.
	M:	Because here there are the highest number of probabilities, and then ...
00:45	D:	Of course
	M:	... the probability is higher.

From this transcript we can see that the students are making sense of the task. Marco is dealing with fractions, he is interested in the procedure. Luca, instead, seems more interested in understanding the overall sense of the activity ("Why don't we first compute how many probabilities are there in all?" 00.36). Both Luca (00:11) and Marco (00:42) come to notice that the highest probability is related to the first room, but seemingly from different standpoints: Luca makes his conclusion based on the fact that room 1 is arrived at by always going left, which has a higher probability than right. We can say that Luca is intuiting the solution rather than computing it. Marco, on the other hand, arrives at the same conclusion much later, by means of computations. Only after considering fractions can he say that room 1 has the highest probability. Davide seems interested in the activity, but his utterances lead us thinking that he is still grasping the sense of the task ("What do we have to compute?" 00:28), and he is struggling to follow Marco's reasoning ("Why?" 00:37).

Marco expressed an instrumental view of mathematics, while Luca showed us a relational view. This divergence of views emerges at the moment Marco repeats, at 00:42, the same utterance Luca said at 00:11—room 1 has the highest probability—as if it is Marco that reached this conclusion for the first time. How is this possible? How is it that Marco does not see Luca's contribution? To begin to understand the answer to this question we need to shift our focus from the strictly literal interactions that we see in the transcript above and move to a more sociocultural analysis of the interaction. To do so we have co-opted the idea of an interactive flowchart.

2 Interactive Flowchart

Interactive flowcharts were introduced by Sfard and Kieran (2001) as a way to capture "two types of speaker's meta-discursive intentions: the wish to react to a previous contribution of a partner or the wish to evoke a response in another interlocutor" (p.58). In this regard, a conversation can be thought of as being comprised of a series of invisible arrows aimed at specific people and/or specific utterances. Sfard and Kieran (2001) developed a coding scheme to make visible these invisible arrows. The scheme follows two basic structures:
- A vertically or diagonally upward arrow is called a reactive arrow and points towards a previous utterance.
- A vertically or diagonally downward arrow is called a proactive arrow and it points towards the person or people from whom a reaction is expected.

Add to this a distinction between arrows that are on task or mathematical in nature (solid) and off-task or non-mathematical in nature (dashed). Sfard and

Kieran (2001) developed this scheme to coded conversations between two people. Ryve (2006) extended this scheme to account for more than two people by assuming that a proactive utterance is meant to address each of the other participants.

Following Ryve (2006) the transcript was coded with arrows (see Table 2). We read the arrows as follows: in the first row, for example, Marco (M) makes a comment to Luca (L) and Davide (D), D responds to M.

What we see when the transcript is coded in this fashion is a very different picture from when we look at the transcript alone. For example, Marco is contributing the most proactive statements (n=7) as opposed to Luca (n=3) or Davide (n=0). Marco and Davide responds to the most number of proactive statements (each n=5) as compared to Luca (n=1 not counting the self-talk as a reaction). Finally, there is a marked difference in the number of proactive statements that each person makes that are reacted to – Marco (n=6), Davide (n=3), and Luca (n=1, not counting the self-talk). We will come back to the issue of self-talk in the section on fictional writing.

The overall impression of this analysis is that Luca is being ignored—although he makes proactive statements, they are not reacted to. Luca had come to the solution very early on, but his classmates seem not to have heard him. The interactive flowchart confirms this, but also shows that it is Luca, more generally, and not just his solution that is ignored – even disregarded. But the video tells a very different story—a story in which Luca is not at all ignored. Communication is a willful activity. Even in groups we want to be heard by certain individuals and we listen to some individuals more than others. Knowledge is emerging from group interaction, but we need suitable tools in order to integrate the individual cognitive/affective realm with a perspective that takes into account the interactions among the group, to see whether and how they can help us in seeing unrepresented relevant aspects.

Table 2: An Interactive Flowchart Coded from the Transcript (L = Luca, D = Davide, M=Marco)

			L	D	M
00:00	M:	To the left two thirds, to the right one third.		o	o
00:01	D:	Yes, I don't remember. (*speaks over M*)	o	o	o
00:03	M:	Then it goes two thirds, two thirds.	o	o	o
00:06	M	Can you give me a pen, please?	o	o	o
00:07	L:	No, let's do the first case, which is the one where it goes always …	o	o	
00:10	M:	… left. You have two thirds here …	o	o	o
00:11	L:	That is the most probable one. (*speaks over M*)	o	o	o
00:13	M:	…and here is one third.	o	o	o
00:15	L:	Should you erase?	o	o	o
00:16	M:	Yes, bravo!	o	o	o
00:17	D:	I'm cute!	o	o	o
00:19	M:	Two thirds and here one third, hence these two thirds…	o	o	
00:21	F:	… they g … they go ….	o	o	o
00:22	M:	Two thirds of two thirds.	o	o	o
00:25	D:	But … but what are you saying? Then no …	o	o	o
00:27	M:	Of these two thirds you should do …	o	o	o
00:28	D:	We have … but what do we have to compute? (*speaks over M*)	o	o	
00:30	L&M:	The probability that the robot will arrive in each one …	o	o	
00:34	M:	of these rooms.	o	o	o
00:35	D:	In the meantime, let's see …	o	o	o
00:36	L:	Why don't we first compute how many probabilities there are in all?	o	o	
00:37	M:	To me this is the room with the highest probability.	o	o	
	D:	Why?	o	o	o
00:42	L:	There are 8 in all.	o	o	o
	M:	Because here there are the highest number of probabilities, and then …	o	o	
00:45	D:	Of course	o	o	o
	M:	… the probability is higher.	o	o	o

3 Interactive Flowchart + Gaze

To truly understand the nature of the interaction in this video we introduce the idea of gaze. That is, we represent, using a new set of arrows, where someone is gazing during each utterance. We use red arrows to represent the speaker and blue arrows to represent non-speakers.

We also introduce a new interlocutor to the interaction – the paper (P) with the problem on it. This paper holds the gaze of the participants at different times of the conversation – so much so that we do not code blue arrows when the students are looking at the paper. Unlike the arrows representing utterances all of the gaze arrows are diagonally downward to represent the passage of time. Table 3 shows the transcript and interactive flowchart overlaid with the gaze arrows.

This flowchart now reveals some interesting aspects of the interaction that the previous version (see table 2) did not show. Whereas the initial interactive flowchart seemed to show that Luca is being ignored the gaze arrows show that this is not at all the case. True, Davide never looks at Luca. Then again Davide doesn't look at anyone – he only looks at the paper when he is speaking. Marco, on the other hand, spends more time looking at Luca (n=6) than at the paper (n=5). At 00:25 Davide is asking a question while gazing at the paper. But Marco is not looking at Davide – he is looking at Luca. Then, while Marco responds to Davide's question at 00:27 he continues to look at Luca. This happens again at 00:34. So, unlike the earlier interactive flowchart, this new flowchart including gazes seems to indicate that Luca is not at all being ignored or disregarded, but rather that he is quite well attended to by Marco.

At the same time Luca only looks at Marco three times. Once at 00:15, then again at 00:25 while Davide is asking a question, and finally 00:36 while Marco is looking at the paper. So, as much as Marco is attending to Luca, Luca is ignoring, maybe even avoiding, Marco. Why is Marco so intent on Luca and why is Luca ignoring Marco? To answer this we need a further modification to the flowchart.

Table 3: Interactive Flowchart with Gaze Arrows

			L	D	M	P
00:00	M:	To the left two thirds, to the right one third.				
00:01	D:	Yes, I don't remember. (*speaks over M*)				
00:03	M:	Then it goes two thirds, two thirds.				
00:06	M	Can you give me a pen, please?				
00:07	L:	No, let's do the first case, which is the one where it goes always …				
00:10	M:	… left. You have two thirds here …				
00:11	L:	That is the most probable one. (*speaks over M*)				
00:13	M:	…and here is one third.				
00:15	L:	Should you erase?				
00:16	M:	Yes, bravo!				
00:17	D:	I'm cute!				
00:19	M:	Two thirds and here one third, hence these two thirds…				
00:21	F:	… they g … they go ….				
00:22	M:	Two thirds of two thirds.				
00:25	D:	But … but what are you saying? Then no …				
00:27	M:	Of these two thirds you should do …				
00:28	D:	We have … but what do we have to compute? (*speaks over M*)				
00:30	L&M:	The probability that the robot will arrive in each one …				
00:34	M:	of these rooms.				
00:35	D:	In the meantime, let's see …				
00:36	L:	Why don't we first compute how many probabilities there are in all?				
00:37	M:	To me this is the room with the highest probability.				
	D:	Why?				
00:42	L:	There are 8 in all.				
	M:	Because here there are the highest number of probabilities, and then …				
00:45	D:	Of course				
	M:	… the probability is higher.				

Table 4 shows the interactive flowchart from 00:25 to 00:45 of the interaction with a slight modification. In this version of the flowchart we have increased the thickness of the gaze arrows according to how intense the gaze is. From this flowchart we can see something interesting happening at 00:25. While Davide is asking the question Luca and Marco are looking at each other. But these are not looks of equal intensity. In the video Marco is clearly more intense in his gaze upon Luca, who, after a while, glances away from Marco. From that moment on

Marco continues to be very intensely focused on Luca. Luca seems to sense this and diverts his gaze from Marco, only looking back at him while Marco is looking at the paper (00:36). Clearly there is an affective aspect to the interaction between Luca and Marco that the flowchart in Table 4 is picking up on. There are emotions, efficacy, will, and motivation in how Luca and Marco are interacting with each other. The gaze intensity arrows can detect this – but to go beyond detection, to really see affect, we need to invoke fictional writing.

Table 4: Interactive Flowchart with Gaze Intensity Arrows

			L	D	M	P
00:25	D:	But … but what are you saying? Then no …				
00:27	M:	Of these two thirds you should do …				
00:28	D:	We have … but what do we have to compute? (*speaks over M*)				
00:30	L&M:	The probability that the robot will arrive in each one …				
00:34	M:	of these rooms.				
00:35	D:	In the meantime, let's see …				
00:36	L:	Why don't we first compute how many probabilities there are in all?				
00:37	M:	To me this is the room with the highest probability.				
	D:	Why?				
00:42	L:	There are 8 in all.				
	M:	Because here there are the highest number of probabilities, and then …				
00:45	D:	Of course				
	M:	… the probability is higher				

4 Fictional Writing

Fictional writing is a technique that can help the researcher to go beyond the external and visible into the students' inner subjective experience (Hannula, 2003). It is a methodology consisting of envisioning the inner monologue of the student, creating likely impressions, and connections that do not exist in the original data. Although this methodology is subjective in nature, it is not wholly so. With the help of good data and extensive analysis the researcher is able to construct inner monologue that are consistent with the empirical data. These inner monologues can help shed light on the students' emotional disposition, attitudes and beliefs about mathematics. In table 5 we present the inner dialogues of Luca, Davide, and Marco (in italics) embedded within the extended interactive flowchart from Tables 3 and 4.

Now we can see a possible explanation for Luca's behavior. Imagining an inner speech, we can see that Luca is feeling a sense of avoidance about fractions that forces him to think on another level—a level that provides him with an overarching view of the task. Any time Marco uses fractions, Luca escapes. At the same time, Marco seems to have a procedural view of mathematics, "mathematics is doing computations", and in fact he is concerned mostly with computations with fractions. For him, the whole sense of the task is to do computations, which provides him with a sense of likely success. Marco's perceived competence in mathematics also frames his emotional disposition towards the activity and the task—a sense of self-confidence and pleasure. Pleasure is provided by this "feeling good" about computations to be done, and conversely this feeling provides Marco with a basis for his perceived competence in doing math.

Davide can be considered Marco's opposite, not on the axis of pleasure (since we can sense that Davide is involved in the activity, he likes it), but on the "I can do" axis. Davide is aware that he is not a good student in math, he "can't do" it, and this shapes his continuous act of prompting Marco's reactions. Davide has a willingness to understand. Marco, meanwhile, pretends to have understood everything and spread his knowledge to his classmates, like a talented, enlightened student.

Luca is avoiding Marco's gazes, which are very intense and firmly directed towards Luca. We see a sort of power struggle between Luca and Marco, where Luca is not prone to concede to Marco. Another power struggle is played between Marco and Davide, but in the end Davide gives in to Marco. The power struggle between Luca and Marco reflects another struggle: the conflict between differing views of mathematics. This conflict is substantial/thick, to the point that the students' gaze to each other but do not listen to each other. Two conflicting views turn out to impede the interaction between Marco and Luca. Fictional writing provides us with the tools to see *why* there is conflict. Marco gazes a lot at Luca, trying to catch his attention and willing him to agree. Because of this, the two students cannot really interact. The students' words, their gestures, their gazes cannot tell us this information—only by evoking the students' inner speech (inferred from a coordination of speech, posture, and gazes) can we sense the students' deep modes of being, their modes of interacting with each other, and their modes of dealing with the task.

Table 5: Inner Monologues (*in italics*)

			L	D	M	P
00:00	M:	To the left two thirds, to the right one third. *I can do it! I know how to do it!*				
00:01	D:	Yes, I don't remember. (*speaks over M*) *I have no idea. Marco is too fast. I need time. But I like this problem, I want to solve it. I really want to solve it, but Marco is too fast.*				
00:03	M:	Then it goes two thirds, two thirds . *I can do it.*				
00:06	M	Can you give me a pen, please?				
00:07	L:	*Marco is confusing me with all these fractions! Why don't we think about the sense of the task?* No, let's do the first case, which is the one where it goes always ...				
00:10	M:	...... left. You have two thirds here ... *I can do it, I'm actually doing it.*				
00:11	L:	*It feels to me ...* That is the most probable one. (*speaks over M*)				
00:13	M:	...and here is one third *I can do it.*				
00:15	L:	Should you erase? *Uh, Marco has to correct what he has written. I have no idea of what he is doing with all these fractions!*				
00:16	M:	Yes, bravo!				
00:17	D:	I'm cute!				
00:19	M:	Two thirds and here one third, hence these two thirds... *Here we go – this is the way to do it.*				
00:21	F:	...they g... the go.... *Maybe I am understanding something.*				

			L	D	M	P
00:22	M:	Two thirds of two thirds.	o	o	o	o
00:25	D:	But ... but what are you saying? Then no ... *I can't follow Marco. he is too fast.*				
	M:	*Davide is too slow. Luca is with me – he is following what I am doing.*				
	L:	*Marco is trying to take over here. He wants me to go his way.*				
00:27	M:	Of these two thirds you should do ... *Wait – Luca is not with me. Come with me, I know what I am doing.*	o	o		
	L:	*I know that Marco is trying to take charge here. But I don't like it. I don't like all these fractions.*				
00:28	D:	We have ... but what do we have to compute? (*speaks over M*)	o	o		
00:30	L&M:	The probability that the robot will arrive in each one ...	o	o		
00:34	M:	of these rooms. *Luca and I are good in math, Davide is not understanding. Luca is like me. Come with me Luca!*	o	o		
00:35	D:	In the meantime, let's see ... *Wait, wait, wait....*	o	o		
00:36	L:	Why don't we first compute how many probabilities there are in all? *Let's try to avoid all these awful fractions.*	o	o		
00:37	M:	To me this is the room with the highest probability. *Wait – Luca is not with me. Come with me, Luca! I am right. Show me that you know I'm right.*	o	o		
	D:	Why? *Oh my goodness! This is new... Marco is too fast!*	o	o		
00:42	L:	There are 8 in all. *This is a better way – and it avoids all these fractions.*	o	o		
	M:	Because here there are the highest number of probabilities, and then ... *Luca! Listen to me! I am doing it right!*	o	o		
00:45	D:	Of course. *Let's trust Marco.*	o	o		
	M:	... the probability is highe *Got it! This is the answer. This is the way to do it.*	o	o		

5 Discussion

To sum up the story told in this paper, we have started from observing the students' utterances from a cognitive perspective, and we have inferred their emotional dispositions with respect to the activity they are engaged in. How knowledge emerges and is shared amongst the group is seen by means of the words the students say. Then, we have looked at the students' interactions, letting the behavioral lens play the main role in the analysis. Adding the students' glances points further towards the issues arising within the embodied mind paradigm: students' gestures, postures and glances are seen as constitutive components of the meaning making process. The students do not only express and communicate their ideas through the movements of their hands, they do not only stare at each other in order to catch the others' attention, but the ideas that emerge from the activity *are* in their gestures and glances—to the point that if we discard these elements as we did at the beginning of the paper we miss many relevant facts.

But this is not the end of our story. Fictional writing provides us with a lens that helps us go deeper inside the students' thoughts and will. Following Hannula's (2003) techniques and intentions, we came to confirm that in order to open a window on the students' inner world it is necessary to repeatedly, patiently, and carefully look at their interactions, their words, and their postures. Knowledge and emotions are distributed throughout the students' hands, eyes, mind, and body, and they are consubstantial to the development of the activity. Gazes give us insights into this inner world and allow us to write a version of the inner monologues of each participant. Other monologues can be constructed from the data just like other conclusions can be extracted from different analyses. Regardless of what monologues result, however, one thing is clear—the interactions between these four students have a turbulent undercurrent of emotions and intentions. The use of interactive flowcharts documenting the verbal interactions and the gazes gives a window into these emotions and intentions. And like opening a window on a wonderful landscape fictional writing helps us to paint the sounds and smells that populate this landscape.

6 References

Hannula, M. (2003). Fictionalising experiences – experiencing through fiction. *For the Learning of Mathematics, 23*(3), 31-37.

Iversen, K., and Nilsson, P. (2005). Students' meaning-making processes of random phenomena in an ICT-environment. In: *Proceedings of the fourth Conference of the European Research in Mathematics Education*, pp. 601-610.

Roth, W.-M., & Radford, L. (2011). A cultural historical perspective on teaching and learning. Rotterdam: Sense Publishers.

Ryve, A. (2006). Making explicit the analysis of students' mathematical discourses – Revisiting a newly developed methodological framework. *Educational Studies in Mathematics, 62*(1-3), 191–209.

Sfard, A. (2001). There is more to discourse than meets the ears: Looking at thinking as communicating to learn more about mathematical learning. *Educational Studies in Mathematics, 46*(1), 13–57.

Sfard, A. & Kieran, C. (2001). Cognition as communication: Rethinking learning-by-talking through multi-faceted analysis of students' mathematical interactions. *Mind, Culture, and Activity, 8*(1), 42-76.

Epistemological Judgments in Mathematics: An Interview Study Regarding the Certainty of Mathematical Knowledge

Benjamin Rott, Timo Leuders, Elmar Stahl
University of Education Freiburg

Content

1 Introduction the Research Project „LeScEd" 228
    1.1   Theoretical Background 228
    1.2   Development and Implementation of the Interviews 231
2 Sample Setting: Certainty of Mathematical Knowledge 231
    2.1   Theoretical Background in the Philosophy of Mathematics: 231
    2.2   Realization of the Interview: 232
3 Initial Results 233
    3.1   Interviewees judging that "mathematical knowledge is certain" 233
    3.2   Interviewees judging that "mathematical knowledge is uncertain" . 234
    3.3   Interpretation using the category of epistemological judgments 235
4 Discussion 236
5 References 236

Abstract

Research on personal epistemology is confronted with theoretical issues as there exist conflicting data regarding its coherence, discipline-relation and context-dependence as well as methodological issues regarding the often used questionnaires to measure epistemological beliefs. We claim that it is necessary to distinguish between relatively stable "epistemological beliefs" and situation-specific "epistemological judgments". In a sequence of interviews with regard to the topic of "certainty of mathematical knowledge", we show that the usual categories used in questionnaires to measure epistemological beliefs have to be differentiated. We argue that epistemological judgments provide a promising framework to interpret the statements of the interviewees.

1 Introduction the Research Project „LeScEd"

Research orientation is a key characteristic of higher education and university education (cf. Tremp & Futter 2012). It is represented in normative frameworks for educational studies such as teacher education (e.g., KMK 2004). Research orientation is characterized as the competence to receive and understand scientific knowledge ("engagement with research") and in addition to think and work scientifically ("engagement in research") (cf. Borg 2010). A development of these competencies is seen as essential to prepare pedagogical and educational professions in understanding science and science communication.

The research project "LeScEd" (an acronym for "Learning the Science of Education"), which is funded by the BMBF[4], is dedicated to examine three facets of research orientation of university students and doctoral candidates:

- Knowledge and mastery of procedures and methods of social sciences;
- Scientific argumentation and communication;
- Epistemological beliefs about the nature of knowledge and knowing.

This article addresses a subproject which studies epistemological beliefs with a special focus on mathematics. The purpose of the subproject is the construction of an instrument to measure beliefs of students with respect to the epistemology of mathematics. In a first step we investigated the assumption that the usual categories used in questionnaires have to be differentiated. We argue that epistemological judgments can be grounded in different beliefs as well as other cognitive arguments (e.g., Stahl 2011), within a considerable range of sophistication.

1.1 Theoretical Background

A person's beliefs are his/her "[p]sychologically held understandings, premises, or propositions about the world that are thought to be true." (Philipp 2007, p. 259) They filter his/her perceptions and direct his/her actions (cf. Philipp 2007). For example, beliefs about mathematics influence the person's mathematical problem solving performance (e.g., Schoenfeld 1992) and his/her acquisition of mathematical knowledge (see Muis 2004, p. 339 ff. for an overview of related studies).

Epistemology is a branch of philosophy dealing with the nature of human knowledge and its justification. Researchers attribute a growing interest in per-

---

[4] Bundesministerium für Bildung und Forschung – Federal Ministry of Education and Research

sonal epistemology development and epistemological beliefs (= beliefs about the nature of knowledge and knowing) (cf. Hofer & Pintrich 1997).

Research on personal epistemology origins in the work of Piaget and Perry and is nowadays part of both psychology and education (cf. Hofer 2000). Whereas early studies modeled personal epistemology as a unidimensional sequence of stages, recent studies consider personal epistemology as "a system of more or less independent epistemological beliefs" (Hofer 2000, p. 379). Hofer and Pintrich (1997) proposed a structure for that system of epistemological beliefs: According to Hofer and Pintrich there are two general areas with two dimensions each. The first area is nature of knowledge (what one believes knowledge is) with the two dimensions certainty of knowledge and simplicity of knowledge; the second area is nature or process of knowing (how one comes to know) with the dimensions source of knowledge and justification of knowledge. Of special interest for this article is the dimension certainty of knowledge which is defined as follows:

> "Certainty of knowledge. The degree to which one sees knowledge as fixed or more fluid appears throughout the research, again with developmentalists likely to see this as a continuum that changes over time, moving from a fixed to a more fluid view. At lower levels, absolute truth exists with certainty. At higher levels, knowledge is tentative and evolving. [...]" (Hofer & Pintrich 1997, S. 119 f.)

A growing amount of psychological research presents relationships between epistemological beliefs and various aspects of learning. It is generally assumed that more sophisticated epistemological beliefs are related to more adequate learning strategies and therefore better learning outcomes (cf. Hofer & Pintrich 1997; Stahl 2011). However, conflicting data exist that cannot be explained with traditional theories about epistemological beliefs (cf. Bromme, Kienhues, & Stahl 2008). Even though most researchers have conceived this construct as general and rather stable, growing empirical evidence showed that epistemological beliefs are less coherent, more discipline-related and more context-dependent than it was hitherto assumed (cf. Hofer 2000).

For example, Muis, Franco, and Gierus (2011) analyzed the epistemological beliefs of students enrolled in a statistics course. They showed that "slight changes in context influence what epistemic beliefs are activated, which can subsequently influence learning." (ibid., p. 516)

Stahl (2011) claims that it is necessary to distinguish between relatively stable epistemological beliefs and situation specific epistemological judgments when examining this construct in more detail. Epistemological judgments are defined

> "[...] as learners' judgments of knowledge claims in relation to their beliefs about the nature of knowledge and knowing. They are generated in dependency of specific scientific information that is judged within a specific learning context. [...] [A]n

epistemological judgment might be a result of the activation of different cognitive elements (like epistemological beliefs, prior knowledge within the discipline, methodological knowledge, and ontological assumptions) that are combined by a learner to make the judgment." (Stahl 2011, p. 38 f.)

Stahl (2011) elaborates these theoretical considerations of a generative nature of epistemological judgments with fictitious examples. Three persons with different backgrounds (content knowledge, methodological knowledge, ontological assumptions, epistemological beliefs, etc.) in physics each judge the claim that the distance between sun and earth is 149.60 million kilometers. In this article we intend to support this assumption by empirical examples.

In mathematics education the terms "personal epistemology" and "epistemological beliefs" are rarely used. Instead, research on this topic is assessed under the construct of beliefs (cf. Muis 2004, p. 322). Muis summarizes several studies dealing with beliefs about mathematics:

> "The majority of research that has examined students' beliefs about mathematics suggests that students at all levels hold nonavailing[5] beliefs. In general, when asked about the certainty of mathematical knowledge, students believe that knowledge is unchanging. The use and existence of mathematics proofs support this notion, and students believe the goal in mathematics problem solving is to find the right answer. [...]" (Muis 2004, p. 330)

Researchers investigating beliefs about mathematics as a discipline deal with opposing perceptions of mathematics: process-orientation versus rule-orientation, dynamic versus static interpretation, formal versus informal discipline, or its applicability (cf. Muis 2004; Grigutsch, Raatz & Törner 1998).

The global intentions of our research project are (a) to identify epistemological beliefs about mathematics as a science and (b) to develop the instruments to do so economically (cf. Muis 2004, p. 354). The research intentions for this paper are (i) to identify epistemological beliefs about mathematics as a science (especially regarding "certainty of knowledge"), and (ii) to provide empirical evidence that supports the theoretical differentiation between epistemological beliefs and epistemological judgments.

---

[5] To avoid a negative connotation, Muis (2004, p. 323 f.) does not use the common labels "naïve – sophisticated" or "inappropriate – appropriate" from psychological and educational research. Instead she suggests to use the labels "nonavailing – availing" for beliefs that are associated with better learning outcomes ("availing"), and for beliefs that have no influence or a negative influence on learning outcomes ("nonavailing").

1.2 Development and Implementation of the Interviews

Because of our global intentions, we chose suitable positions from the philosophy of mathematics (e.g., about the ontology of mathematical objects) and started to design a manual for semi-structured interviews with the long-term goal to develop an adaptive, web-based questionnaire to collect data about according beliefs. The aim of the interviews is to examine the idea of a generative nature of beliefs in more detail.

To get more insight into our subjects' beliefs, we did not just ask general questions with philosophical orientation but presented quotes of representatives of opposing epistemological positions and had our subjects relate themselves thereto. Afterwards, we intervened with information contrary to the subjects' positions to further identify their lines of reasoning.

During the first phase of data collection, we optimized our selection of quotes as well as our interview questions and developed additional interventions for the various subjects' positions and reasons. This can be seen as an application of Grounded Theory (cf. Strauss & Corbin 1996) which also postulates that the research design can be developed further with respect to successively analyzed data. As topics for the interviews we chose different key questions on the epistemology of mathematics as a science. In the following we chose to present – as an example from the larger body of data we collected – a single setting which deals with the topic of certainty of mathematical knowledge.

2 Sample Setting: Certainty of Mathematical Knowledge

2.1 Theoretical Background in the Philosophy of Mathematics:

Mathematical knowledge is regarded as certain since antiquity, because of formal proofs and deductive reasoning with respect to valid rules and axioms (cf. Heintz 2000, p. 52 ff.; Hoffmann 2011, p. 1 ff.) But this belief was shaken several times during the history of mathematics: (i) It is impossible to justify the axioms that theorems rely on and the discovery of non-Euclidean geometries has shown that different determinations can lead to divergent mathematics. (ii) The finding of contradictory derivations from axioms (Russell's paradox) led to the attempt of establishing formal rules of derivation by D. Hilbert but was doomed to failure because of Gödel's incompleteness theorems in 1931 (cf. Hoffmann 2011, p. 52 ff.). (iii) Proofs of mathematical theorems can be inaccurate or even incorrect and the review process of publishing magazines cannot guarantee identifying all weak spots. Often, mathematical work is so specialized that only a handful of experts is able to comprehend it and the history of mathematics is full of examples of accepted proofs that were discovered to be wrong years after their publication. (iv) Finally, a growing number of mathematical results is achieved with the help of computers and no living mathematician is able to verify them without trusting the machines as well as hoping for error-free hard- and software (cf. Borwein & Devlin 2011, p. 8 ff.).

These aspects all relate to the topic of certainty of mathematics. Interviewees can relate to these aspects in different ways when arguing about their individual judgment on the certainty of mathematics.

Table 1 Starting positions for "Certainty of Mathematical Knowledge".

Mathematical knowledge is certain	Mathematical knowledge is uncertain
"In mathematics knowledge is valid forever. A theorem is never incorrect. In contrast to all other sciences, knowledge is accumulated in mathematics. [...]	"The issue is [...] whether mathematicians can always be absolutely confident of the truth of certain complex mathematical results [...].
It is impossible, that a theorem that was proven correctly will be wrong from a future point of view. Each theorem is for eternity."	With regard to some very complex issues, truth in mathematics is that for which the vast majority of the community believes it has compelling arguments. And such truth may be fallible.
(Albrecht Beutelspacher) [2001, p. 235; translated by the first author]	Serious mistakes are relatively rare, of course."
	(Alan H. Schoenfeld) [1994, p. 58 f.]

2.2 *Realization of the Interview:*

We confronted to our subjects with two quotes (see Table 1) and invited them to answer the following prompt: "These are two positions of mathematicians regarding the certainty of mathematical knowledge. With which position can you identify yourself? Please give reasons for your answer."

Further questions were: "Can you explain your position on the basis of your mathematical experience?" "Please compare the certainty of mathematical knowledge to that of other sciences, for example to physical, linguistic, or educational knowledge."

If a subject settled on "math knowledge is certain", we confronted him/her with the story of a false proof of the four color theorem by A. Kempe in 1879 that was accepted by the community of mathematicians and which was shown to be false by P. Heawood not earlier than 11 years later (e.g., Wilson 2002). If a subject thought that "math knowledge is uncertain", we asked whether a theorem like the Pythagorean one could be uncertain as there are hundreds of proofs, countless validations and practical applications like in masonry.

So far, the first author interviewed 10 pre-service teachers of mathematics (students at the University of Education Freiburg), 2 in-service teachers of mathematics, 2 professional mathematicians and 2 professors of mathematics. Below you'll find a selected sample of these interviews.

3 Initial Results

Our initial results with respect to the area of "certainty of math knowledge" are twofold: Firstly, we present two different lines of reasoning each for "certain" and "uncertain" to point out what arguments we found empirically to support these positions. Secondly, we show that subjects who support the same position and would (and actually did) check the same boxes for according questions in a typical beliefs questionnaire can do so for differing reasons. This supports our argument for the theoretical introduction of *epistemological judgments*.

3.1 Interviewees judging that "mathematical knowledge is certain"

1) T.W. is a pre-service teacher in his second year at the University of Education in Freiburg. For him, mathematical knowledge is certain, "the first quote of Beutelspacher is more likely correct in my view." He says that he thinks of proofs as inevitable and irrefutable. And he adds: "How can there possibly be errors in mathematics?"

Confronted with the historical episode of the four color theorem, T.W. admits "Of course, there can be errors, [...] but it got proven eventually, didn't it?" When asked why he was so sure about the certainty of mathematical knowledge, T.W. mentioned the Pythagorean Theorem as an example of a theorem which is inevitable for him, which has several hundred proofs, and which will not change in the next 10, 100, or 1000 years. He concludes with: "Hopefully. Otherwise, my fundamental conception would be destroyed."

2) A.R., a mathematician who just finished his diploma at the University of Oldenburg, considers mathematical knowledge for certain: "I identify myself definitely with Albrecht Beutelspacher." A.R. adds that errors are possible, but these would be the errors of mathematicians but not of mathematics itself. "Humans are fallible. [...] There might be errors in proofs which are accepted by many people. [...] But when a theorem is proved correctly from the axioms by formal rules of derivation then it will last for eternity." As an example for human fallibility A.R. refers to Andrew Wiles' proof of Fermat's Last Theorem. This was regarded as proven for a short time, then rejected and republished after some years. It might take another several years until the methods Wiles used pass over to the common mathematical knowledge, but A.R. believes that there will be a time when this theorem and its proof will have been checked thoroughly and will have been finally accepted as certain.

The quote of Schoenfeld might go well with great mathematical puzzles like Riemann's Conjecture but otherwise, it does not describe A.R.'s view of mathematical knowledge. In comparison, other scientific disciplines are dependent on tests and laboratory experiments which results in their knowledge being uncertain. In contrast, mathematical knowledge is reducible to basic elements, the axioms, and to logical conclusions, which makes it certain.

3.2   Interviewees judging that "mathematical knowledge is uncertain"

3) B.G. is a pre-service teacher who just finished her degree at the University of Education in Freiburg. She thinks that mathematical knowledge is uncertain, because "for me, there is always the possibility that someone figures out that something is not quite correct. A theorem might be proven and checked but there is always the possibility of finding an aspect that it may not be correct." She generally would not agree to any statement regarding "ever" or "never".

Asked if there is a counter-example for the Pythagorean Theorem she responds that she is not able to come up with any, but there might be others with a better mathematical background who could. The interviewer wanted to know if she was certain of the consequences of her position. The logical construction of mathematics might collapse if basic findings like the Pythagorean Theorem or Complete Induction were not certain. B.G. responded with "I know of the consequence and I'm fine with it. [...] This is no problem for me. [...] But the possibility for this to happen is very, very small."

She states that in comparison to other scientific disciplines, knowledge in mathematics is very certain, but some uncertainty remains.

4) S.W. is a mathematics professor at the University of Hanover for several years. In his view, Beutelspacher holds a Platonic view which he cannot agree to. He says that he does not believe in a mathematical realm with eternal conceptions that exists outside the human sphere. S.W. describes in detail that he can think of basically two arguments for mathematical knowledge being uncertain: Firstly, mathematicians are fallible and errors can occur during proving and reviewing. But this is not the main point, because the community is very careful and all but maybe the most complex things are thought through very thoroughly and therefore very certain. Secondly, this is the crucial point according to S.W., mathematical knowledge cannot be definitely certain because that would imply an infallible system of rules with an otherworldly justification. Mathematics would need a justification outside of the human sphere and outside of the mathematical discourse, a realm that could be observed and described. S.W. concludes: "That there is such a realm, such a sphere, I am very skeptical about it."

Asked whether theorems like the Pythagorean one are not sure, S.W. answers: "This theorem cannot be certain because it is unclear what certainty means in this context." He says that the Pythagorean Theorem is an innermathematical theorem that the community of mathematicians considers true under certain axiomatic assumptions. But this does not mean that it would be true if there were no humans or the universe came to an end, because of the missing mathematical realm that would justify such eternal truth.

In comparison to other scientific disciplines, S.W. states that mathematical knowledge is more certain, because the other sciences have the same problem of a missing justification as well as additional disturbances in the form of assumptions, hypothesis and doctrines. "Mathematics is more rigorous and so to speak more pure and therefore more certain in a sense."

3.3 Interpretation using the category of epistemological judgments

Both T.W. and A.R. answered the knowledge claim whether mathematical knowledge is certain or uncertain in the same way and both checked the box "mathematics is very certain" in a questionnaire (CAEB, Stahl & Bromme 2007) they completed previously to the interview. Within a questionnaire study this would contribute to positioning them on a belief scale with respect to certainty.

But actually they have shown a substantially different argumentation in those interviews due to their background. Whereas T.W. could only refer to simple examples such as the Pythagorean Theorem, A.R. was able to activate more content knowledge in the form of Andrew Wiles' proof as well as the Riemann Conjecture. Additionally, A.R. did argue with mathematical axioms and rules of derivation, whereas T.W. solely relied on his "fundamental conceptions". A.R. was conscious about possible errors in mathematical proofs but was able to integrate this into his beliefs. T.W., on the other hand, was not aware of this fact up to the intervention and did not use this piece of information for his argumentation ("it got proven eventually, didn't it?").

The same is true for B.G. and S.W. who both supported the position that "mathematical knowledge is uncertain". B.G. could only rely on fundamental conceptions ("I generally do not agree to statements referring to 'ever' or 'never'."). On the other hand, S.W. could not only refer to his content knowledge about the fallibility of the mathematical review process, but also to his ontological knowledge regarding Platonism to support his arguments.

This empirical data supports the theoretical claim of Stahl (2011, p. 49):

> "In a questionnaire with rating scales, [these] persons would give the same answer. However, the conclusion that their responses are an expression for comparable epistemological beliefs would be wrong. Their epistemological judgments are built on different cognitive elements to evaluate the knowledge claim." (Stahl 2011)

4 Discussion

The evaluations of the interviews show the breadth of arguments for the positions of "mathematical knowledge is certain / uncertain". There are more or less reflected representatives of both statements which is somewhat surprising in relation to results from research on epistemological beliefs. For the dimension *certainty of knowledge* more "sophistication" is seen as less belief in truth with certainty (cf. Hofer & Pintrich 1997; Hofer 2000). But the example of A.R. shows that this position can be held in a reflected way (which is revealed by the way he judges the certainty of knowledge of other scientific disciplines).

The evaluations of the interviews also show the gain of the theoretical introduction of *epistemological judgments*. Persons that hold the same position regarding the certainty of mathematical knowledge can do so with differing backgrounds. A traditional beliefs questionnaire would not be able to detect or explain those differences. Therefore it seems doubtful to rely on instruments that measure epistemological beliefs as a locus on a scale. It should be necessary to take into account different strands of argumentation and different backgrounds. The concept of "epistemological judgment" can be a promising starting point for developing instruments that can capture such important differences.

Future prospects include developing an adaptive, web-based questionnaire to measure epistemological judgments and beliefs. A first pilot study was conducted in July 2013 with 45 university students; a second one is scheduled for the 2013/14 winter term.

5 References

Beutelspacher, Albrecht (2001). *Pasta all'infinito – Meine italienische Reise in die Mathematik*. [*My Italian Journey into Mathematics*] München: dtv.

Borg, Simon (2010). Language Teacher Research Engagement. In *Language Teaching, 43 (4)*, pp. 391 – 429.

Borwein, Jonathan & Devlin, Keith (2011). *Experimentelle Mathematik – Eine beispielorientierte Einführung*. [*Experimental Mathematics – An Introduction*] Heidelberg: Spektrum.

Bromme, Rainer; Kienhues, Dorothe; & Stahl, Elmar (2008). Knowledge and epistemological beliefs: An intimate but complicate relationship. In Khine, Myint Swe (Ed.). *Knowing, knowledge and beliefs. Epistemological studies across diverse cultures*. New York: Springer, pp. 423 – 441.

Heintz, Bettina (2000). *Die Innenwelt der Mathematik*. [*The Inner World of Mathematics*] Wien: Springer.

Hofer, Barbara K. (2000). Dimensionality and Disciplinary Differences in Personal Epistemology. In *Contemporary Educational Psychology 25 (2000)*, pp. 378 – 405.

Hofer, Barbara K. & Pintrich, Paul R. (1997). The Development of epistemological Theories: Beliefs About Knowledge and Knowing and Their Relation to Learning. In *Review of Educational Research 1997, Vol. 67, No. 1*, pp. 88 – 140.

Hoffmann, Dirk (2011). *Grenzen der Mathematik*. [*Limits of Mathematics*] Heidelberg: Spektrum.

Grigutsch, Stefan; Raatz, Ulrich; & Törner, Günter (1998). Einstellungen gegenüber Mathematik bei Mathematiklehrern. [*Attitudes of Mathematics Teachers towards Mathamatics*] In *Journal für Mathematik-Didaktik 19 (98) 1*, pp. 3 – 45.

KMK – Kultusministerkonferenz (2004). Standards für die Lehrerbildung: Bildungswissenschaften. Beschluss der Kultusministerkonferenz. [Standards for Teacher Education: Educational Sciences. Resolution of the Standing Conference of Education Ministers.] (10.05.2013): http://www.kmk.org/fileadmin/veroeffentlichungen_beschluesse/2004/2004_12_16-Standards-Lehrerbildung.pdf

Muis, Krista R. (2004). Personal Epistemology and Mathematics: A Critical Review and Synthesis of Research. In *Review of Educational Research 2004, 74, No. 3*, pp. 317 – 377.

Muis, Krista R.; Franco, Gina M.; & Gierus, Bogusia (2011). Examining Epistemic Beliefs Across Conceptual and Procedural Knowledge in Statistics. In *ZDM Mathematics Education (2011), 43*, pp. 507 – 519.

Philipp, Randolph A. (2007). Mathematics Teachers' Beliefs and Affect (Chapter 7). In Lester, Frank K. (Ed.). *Second Handbook of Research on Mathematics Teaching and Learning*, pp. 257 – 315.

Schoenfeld, Alan H. (1994). Reflections on Doing and Teaching Mathematics (Chapter 3). In Schoenfeld, Alan H. (Ed.). *Mathematical Thinking and Problem Solving.* Hillsdale, NJ: Lawrence Erlbaum Associates, pp. 53 – 69.

Schoenfeld, Alan H. (1992). Learning to Think Mathematically. In Grouws, Douglas A. (Ed.). *Handbook for Research on Mathematics Teaching and Learning.* New York: MacMillan, pp. 334 – 370.

Stahl, Elmar & Bromme, Rainer (2007). The CAEB: An instrument for measuring connotative aspects of epistemological beliefs. In *Learning and Instruction 17 (2007),* pp. 773 – 785.

Stahl, Elmar (2011). The Generative Nature of Epistemological Judgments: focusing on Interactions Instead of Elements to Understand the Relationship Between Epistemological Beliefs and Cognitive Flexibility (Chapter 3). In Elen, Jan; Stahl, Elmar; Bromme, Rainer & Clarebout, Geraldine (Eds.). *Links Between Beliefs and Cognitive Flexibility – Lessons Learned.* Dordrecht: Springer, pp. 37 – 60.

Strauss, Anselm L.; Corbin, Juliet M. (1996). *Grounded Theory: Grundlagen Qualitativer Sozialforschung.* [*Grounded Theory: Basics of Qualitative Social Research.*] Weinheim: Beltz.

Tremp, Peter & Futter, Kathrin (2012). Forschungsorientierung in der Lehre. Curriculare Leitlinie und studentische Wahrnehmungen. [Research Orientation in Teaching. Curricular Guidelines and Students' Perception.] In Brinker, Tobina & Tremp, Peter (Eds.). *Einführung in die Studiengangentwicklung.* Bielefeld: Bertelsmann, pp. 69 – 79.

Wilson, Robin (2002). *Four Colors Suffice.* Princeton University Press.

Authors

Chiara Andrà
University of Torino, Torino, Italy
chiara.andra@gmail.com

Benita Berg
Mälardalen University, Västerås, Sweden
benita.berg@mdh.se

Carola Bernack-Schüler
University of Education Freiburg, Freiburg, Germany
bernack@ph-freiburg.de

Marc Bosse
University of Duisburg-Essen, Essen, Germany
marc.bosse@uni-due.de

Katinka Bräunling
University of Education Freiburg, Freiburg, Germany
katinka.braeunling@ph-freiburg.de

Andreas Ebbelind
Linnaeus University, Växjö, Sweden
andreas.ebbelind@lnu.se

Andreas Eichler
University of Kassel, Kassel, Germany
eichler@mathematik.uni-kassel.de

Ralf Erens
University of Education Freiburg, Freiburg, Germany
erens@ph-freiburg.de

Boris Girnat
University of Applied Sciences and Arts Northwestern Switzerland, Basel, Switzerland
boris.girnat@fhnw.ch

Lenni Haapasalo
University of Eastern Finland, Joensuu, Finland
lenni.haapasalo@uef.fi

Markku S.Hannula
University of Helsinki, Helsinki, Finland
markku.hannula@helsinki.fi

Kirsti Hemmi
Mälardalen University, Västerås, Sweden
kirsti.hemmi@mdh.se

Lars Holzäpfel
University of Education Freiburg, Freiburg, Germany
lars.holzaepfel@ph-freiburg.de

Martin Karlberg
Uppsala University, Uppsala, Sweden
martin.karlberg@edu.uu.se

Igor Kontorovich
Technion – Israel Institute of Technology, Haifa, Israel
ik@tx.technion.ac.il

Authors

Timo Leuders
University of Education Freiburg, Freiburg, Germany
leuders@ph-freiburg.de

Peter Liljedahl
Simon Fraser University, Burnaby, Canada
liljedahl@sfu.ca

Liisa Näveri
University of Helsinki, Helsinki, Finland
liisa.naveri@helsinki.fi

Susanna Oksanen
University of Helsinki, Helsinki, Finland
susanna.oksanen@helsinki.fi

Hanna Palmér
Linnaeus University, Växjö, Sweden
hanna.palmer@lnu.se

Ioannis Papadopoulos
Aristotle University of Thessaloniki, Thessanoliki, Greece
ypapadop@eled.auth.gr

Erkki Pehkonen
University of Helsinki, Helsinki, Finland
erkki.pehkonen@helsinki.fi

Päivi Portaankorva-Koivisto
University of Helsinki, Helsinki, Finland
paivi.portaankorva-koivisto@helsinki.fi

Benjamin Rott
University of Duisburg-Essen, Essen, Germany
benjamin.rott@uni-due.de

Laia Saló i Nevado
University of Helsinki, Helsinki, Finland
laia.salo@helsinki.fi

Elmar Stahl
University of Education Freiburg, Freiburg, Germany
elmar.stahl@ph-freiburg.de

Lovisa Sumpter
Dalarna University, Falun, Sweden
lsm@du.se

Maria Sundberg
Dalarna University, Falun, Sweden
msu@du.se

Günter Törner
University of Duisburg-Essen, Duisburg, Germany
guenter.toerner@uni-due.de

Leonor Varas
Universidad de Chile, Santiago de Chile, Chile
mlvaras@dim.uchile.cl

Bernd Zimmermann
Friedrich Schiller University of Jena, Jena, Germany
bernd.zimmermann@uni-jena.de

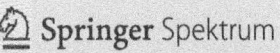

Springer Spektrum Research
Forschung, die sich sehen lässt

Ausgezeichnete Wissenschaft

Werden Sie AutorIn!

Sie möchten die Ergebnisse Ihrer Forschung in Buchform veröffentlichen?

Seien Sie es sich wert. Publizieren Sie Ihre Forschungsergebnisse bei Springer Spektrum, dem führenden Verlag für klassische und digitale Lehr- und Fachmedien im Bereich Naturwissenschaft I Mathematik im deutschsprachigen Raum.
Unser Programm Springer Spektrum Research steht für exzellente Abschlussarbeiten sowie ausgezeichnete Dissertationen und Habilitationsschriften rund um die Themen Astronomie, Biologie, Chemie, Geowissenschaften, Mathematik und Physik.
Renommierte HerausgeberInnen namhafter Schriftenreihen bürgen für die Qualität unserer Publikationen. Profitieren Sie von der Reputation eines ausgezeichneten Verlagsprogramms und nutzen Sie die Vertriebsleistungen einer internationalen Verlagsgruppe für Wissenschafts- und Fachliteratur.

Ihre Vorteile:

Lektorat:
- Auswahl und Begutachtung der Manuskripte
- Beratung in Fragen der Textgestaltung
- Sorgfältige Durchsicht vor Drucklegung
- Beratung bei Titelformulierung und Umschlagtexten

Marketing:
- Modernes und markantes Layout
- E-Mail Newsletter, Flyer, Kataloge, Rezensionsversand, Präsenz des Verlags auf Tagungen
- Digital Visibility, hohe Zugriffszahlen und E-Book Verfügbarkeit weltweit

Herstellung und Vertrieb:
- Kurze Produktionszyklen
- Integration Ihres Werkes in SpringerLink
- Datenaufbereitung für alle digitalen Vertriebswege von Springer Science+Business Media

Sie möchten mehr über Ihre Publikation bei Springer Spektrum Research wissen? Kontaktieren Sie uns.

Marta Schmidt
Springer Spektrum | Springer Fachmedien
Wiesbaden GmbH
Lektorin Research
Tel. +49 (0)611.7878-237
marta.schmidt@springer.com

Springer Spektrum I Springer Fachmedien Wiesbaden GmbH

The manufacturer's authorised representative in the EU is Springer Nature Customer Service Centre GmbH, Europaplatz 3, 69115 Heidelberg, Germany. If you have any concerns regarding our products, please contact ProductSafety@springernature.com

Printed and bound by CPI Group (UK) Ltd, Croydon, CR0 4YY

02076674-0006